THAT CAN'T BE RIGHT!

HOW TO ORDER THIS BOOK

BY PHONE: 800-233-9936 or 717-291-5609, 8AM–5PM Eastern Time

BY FAX: 717-295-4538

BY MAIL: Order Department
Technomic Publishing Company, Inc.
851 New Holland Avenue, Box 3535
Lancaster, PA 17604, U.S.A.

BY CREDIT CARD: American Express, VISA, MasterCard

BY WWW SITE: http://www.techpub.com

PERMISSION TO PHOTOCOPY–POLICY STATEMENT

Authorization to photocopy items for internal or personal use, or the internal or personal use of specific clients, is granted by Technomic Publishing Co., Inc. provided that the base fee of US $3.00 per copy, plus US $ .25 per page is paid directly to Copyright Clearance Center, 222 Rosewood Drive, Danvers, MA 01923, USA. For those organizations that have been granted a photocopy license by CCC, a separate system of payment has been arranged. The fee code for users of the Transactional Reporting Service is 1-56676/99 $5.00 + $ .25.

THAT CAN'T BE RIGHT!

Using Counterintuitive Math Problems

Nelson John Maylone, BA, MA

LANCASTER · BASEL

That Can't Be Right!
a**TECHNOMIC**publication

Technomic Publishing Company, Inc.
851 New Holland Avenue, Box 3535
Lancaster, Pennsylvania 17604 U.S.A.

Copyright ©1999 by Technomic Publishing Company, Inc.
All rights reserved

No part of this publication may be reproduced, stored in a
retrieval system, or transmitted, in any form or by any means,
electronic, mechanical, photocopying, recording, or otherwise,
without the prior written permission of the publisher.

Printed in the United States of America
10 9 8 7 6 5 4 3 2 1

Main entry under title:
   That Can't Be Right! Using Counterintuitive Math Problems

A Technomic Publishing Company book
Bibliography: p.
Includes index p. 127

Library of Congress Catalog Card No. 98-86275
ISBN No. 1-56676-676-1

To Richard Macon and to Robert Chandler, who affected more lives than they can know.

CONTENTS

Preface xi

Acknowledgements xiii

Chapter 1: Introduction 1
   What This Book Is About 1
   About "Counterintuitive" Problems 4
   Characteristics of Adolescents 4
   The Effective Teacher 6
   Conclusion 18
   References 19

Chapter 2: Counterintuitive Problems 21
   #1: The Busy Lawyers 21
   #2: Why Isn't All the Money Gone? 22
   #3: What's the Cost? 25
   #4: The Pen and Pencil Set 25
   #5: What Does the Clock Cost? 26
   #6: The Ditch Diggers 27
   #7: The Raffle 27
   #8: That's Not Enough Information! Part 1 28
   #9: That's Not Enough Information! Part 2 30
   #10: Rolling the Block 32
   #11: The Ribbon around the World 33
   #12: The Careless Cat 34
   #13: Take Your Medicine 35
   #14: High Stakes 35
   #15: Cut-Up 36
   #16: Need a Lift? 36
   #17: The Runner 37

#18: The Hungry Bookworm 37
#19: The Bulb on the Bicycle Tire 38
#20: The Magic Egg Basket 39
#21: Birthday Match 40
#22: A Month of Sundays 41
#23: How Long? 42
#24: Kount the Korn Kernels 44
#25: Socks in the Dark 46
#26: They Can't Be Equal! 46
#27: The Last Day 48
#28: I'm My Own Grandmother! 48
#29: The Three Doors 50
#30: Odds on Two Boys 53

Chapter 3: Challenges and Puzzles   57

#31: Truthteller or Liar? 57
#32: Two Trains, One Unfortunate Fly 58
#33: Let's Play Weatherperson 59
#34: That's Just Perfect 60
#35: Complete the Sequence 61
#36: Convert if You Can 62
#37: The Perfect Solitaire Deck 63
#38: Rotate the Triangle 63
#39: The Shrinking Dollar Bill 65
#40: The Snobby Neighbors 65
#41: It Looks Simple 67
#42: Connect the Dots 69
#43: Two American Coins 70
#44: Into the Forest 70
#45: The Frozen Clock 71
#46: Count the Rectangles 71
#47: The Counterfeit Coins 73
#48: A Streetcar Named Deception 75
#49: Four Gallons, Please 76
#50: Double My Money 76
#51: Palindromic Fun 77

Chapter 4: Demonstrations, Games, and Activities   79

#52: 7 × 13 = 28 79
#53: The "All Answers Are Numbers" Game 82

#54: Complete the Cliché	85
#55: Are You Inferring . . . ?	86
#56: An Infinity of Infinities	87
#57: You Are My Density	88
#58: The Spreading Virus (AKA the Spreading Secrets)	89
#59: How'd He Do That? Part I. The Kangaroos-in-Denmark Trick	91
#60: How'd He Do That? Part II. The 20-to-50 Trick	92

Chapter 5: Investigations — 95

#61: Zeno's Paradoxes	95
#62: How About a Nice Piece of Pi?	97
#63: What's a Mortgage? (or, "I Have to Pay Back How Much?")	99
#64: A Few Notes	100
#65: Boots in Hobble Junction	103
#66: ESP?	105
#67: Roll the Dice—A Probability Investigation	107

Appendix A: Problem-Solving Strategies — 111
Good Problem-Solving Techniques 111

Appendix B: Multiple Intelligences Theory — 115
Examples 116

Appendix C: Eighteen Ideas for Middle-Grade Mathematics Classrooms (a Potpourri) — 117

Index — 127

About the Author — 131

PREFACE

MANY PARENTS OF middle schoolers have been heard to say "I don't like math, and I don't blame my child for not liking it either." Even some teachers occasionally denigrate the study of mathematics with comments such as "I know this is no fun, but you *need* to study math." Although this book purports to be about puzzles and logic and problem solving, its really about breaking that cycle of math negativity.

The problems and activities contained in *That Can't Be Right!* are designed to simultaneously engage and challenge middle-grade students; indeed, the fact that the learners need to wrestle with so many of the items is precisely what tends to engage them. This book was written with the presumption that, contrary to popular belief, adolescents generally enjoy difficult but rewarding work. (Experienced teachers know that adolescent student protests must always be taken with a grain of salt!)

Many critics of American public education decry a supposed dramatic and precipitous decline in students' mathematics "test scores," but careful examination of the facts rarely supports this concern. What is troubling, however, is the apparent loss of a sense of *number* in America. What *are* we to make of the seventeen-year old fast food clerk who cannot mentally figure 6% sales tax on a $2 purchase when the electronic cash register is down? That clerk does not need to remember any formula from math class; he needs to have "numeracy," as Dr. John Paulos of Temple University calls it; a sense of comfort regarding numbers and their relationships.

This book attempts to allow students (and their teachers) to "dive into" thorny math and logic problems, and to enjoy the dive by lowering the risk generally associated with such efforts. The reader will notice over and over again these two themes:

- encouraging student brainstorming (one of the rules of which is that no criticism of comments is allowed)

- fostering divergent student thinking

"Fostering divergent thinking" should not be seen as "anything goes," or as "any and all answers are correct." Students' minds are wonderful and complex things, and giving them some mental room to arrive at answers (correct answers) is essential for promoting numeracy. More importantly, perhaps, when teachers take this more open and accepting approach to the teaching of mathematics, students are more likely to enjoy their studies, thereby helping to break that math–negativity cycle.

It's hoped that teachers and students alike find *That Can't Be Right!* to be a stimulating addition to middle-grade mathematics classrooms, and that it will prompt teachers to continue making their classes interesting, challenging, and in tune with the needs of adolescent students.

ACKNOWLEDGEMENTS

THE AUTHOR IS grateful to mathematics teacher Dick Sietz of Anderson High School in Southgate, Michigan, for suggesting Problem #2, and to mathematics teacher Lisa Rombes of Sietz Middle School, Riverview Community Schools, Riverview, Michigan, for supplying the computer program that accompanies Problem #29, and to *Parade Magazine* and Marilyn vos Savant, whose "Ask Marilyn" columns inspired Problems #29 and 30, and to his daughter Susan for contributing Problem #59. Many other problems are old or ancient "classics" whose origins are unclear, except where noted. All other problems were developed by the author, or are believed to be public domain items.

The author is also very appreciative to Dylan Lapinski for the illustrations used throughout the book.

CHAPTER 1

Introduction

WHAT THIS BOOK IS ABOUT

THIS BOOK IS designed to provide middle-grade mathematics teachers with ideas for enlivening instruction and for engaging students.

"Today we will be learning how to divide unlike fractions, whole and mixed numbers. Please listen to my explanation carefully. If you do, you will be able to correctly answer questions 1–36 found on pages 109 and 110 in your textbooks. Let me now begin. When dividing fractions . . ."

Does the lesson introduction above, provided by a middle-grade mathematics teacher, sound like one which is likely to pique the interest of the typical adolescent? Will it help to engage most students in the lesson? Most importantly, will this type of introduction promote student learning?

The answer to each question is, of course, *no*. Furthermore, if that lesson introduction is any indication of things to come, students in that classroom are in for an exceptionally dry presentation, and the traditional view that much of mathematics is boring and irrelevant will, unfortunately, be confirmed in students' minds.

The urgent need for continuing a revolution in mathematics education in America was well-framed by Dr. Kathe Rasch in her article, "The Imperative for Quality Middle School Mathematics Curriculum and Instruction" (*Midpoints*, National Middle School Association, 1994):

> Without changes in the middle school curriculum and instruction, students will perceive mathematics as dull, repetitive and unconnected to the rest of their lives. Tell, show and drill have been the steadfast, traditional methods of mathematics teaching. Increased knowledge about the cognitive, social and emotional development of middle schoolers indicates that teachers must begin to reconceptualize teaching and learning in mathematics so that students leave middle school empowered and confident about their ability to reason mathematically. (p. 2)

Since their introduction in the late 1980s, The Curriculum and Evaluation Standards of the National Council of Teachers of Mathematics (NCTM) have sparked a revolution in curriculum design, in textbook design, and, to a lesser extent, in the evaluation of students. The Standards encourage, among other things,

- problem solving in context
- providing opportunities for students to discuss mathematics
- actively involving students in exploring, conjecturing, analyzing, and applying mathematics
- using instructional techniques that include group work, discussion, making presentations, conducting experiments, and working with manipulatives

Also, from the NCTM Curriculum and Evaluation Standards for Mathematics: "Learning should engage students both intellectually and physically. They must become active learners, challenged to apply their prior knowledge in new and . . . more difficult situations."

The Standards have moved many teachers to embrace improved instructional practices. Nonetheless, promotion of student engagement in the mathematics classroom remains weak throughout the country.

The key theses of *That Can't Be Right!* are these:

(1) Counterintuitive and other high-allure mathematics questions used in a problem-solving context will help to keep adolescent students interested in mathematics.

(2) Teaching that incorporates students' intuitive solutions can increase learning, especially when combined with opportunities for student interaction and discussion.

(3) Using the types of problems found in this book in mathematics classes opens doors to wonderful interdisciplinary opportunities, particularly in the area of language arts.

(4) As students are enthusiastically engaged, discipline problems will occur less frequently.

As most any middle-school teacher will confirm, adolescents enjoy debating in general; most of all, they enjoy debating adults. Debate, along with its tamer cousin, discussion, are excellent methods for engaging middle-grade mathematics students.

Many research findings support the idea that students who are actively engaged in their mathematics instruction are higher achievers. The Center for Research on Effective Schooling for Disadvantaged Students, The Johns Hopkins University, finds that middle school students who are exposed to problem-solving activities in math class have higher math scores and less fear about asking questions than those whose instruction largely emphasized drill and practice. Students in problem-solving-oriented classes also tend not to be bored in class, to complete their homework, and to state that their schoolwork will prove valuable in the future (from Epstein and MacIver, 1992).

A report from the National Center for Research on Teacher Learning, Michigan State University (Kirsner and Bethell, 1992) states that active learning techniques promote meaningful thought in mathematics—even among students who appear to be disinterested in the subject. Researchers taught math to students who had relearned low-level computational skills year after year. After curriculum and instruction were reorganized to reflect the recommendations of the NCTM, those previously uninvolved students became interested in mathematics.

Discussion and debate promote new thinking; they reinforce learning and offer important opportunities for social erudition. They can help teach youngsters how to listen, how to be patient (a characteristic not often seen as typically adolescent), how to respond appropriately to others' comments and opinions, how to be concise, and how to be persuasive.

In addition to promotion of student discussion and debate in the middle-grade mathematics classroom, this book emphasizes the importance of student writing as well. Writing about mathematics remains, unfortunately, a neglected activity in many classrooms. Writing, like discussion, reinforces learning. It also provides students with a wonderful means of self-expression. Ideas for promoting writing in mathematics classes are provided throughout the book.

Good teachers know that student assessment is much more than paper-and-pencil tests. Meaningful and authentic assessment includes observing students as they tackle performance tasks, the assembly of portfolio materials, conferencing with students, and furnishing students with opportunities to write and speak in all curricular areas (including mathematics). Careful observation of youngsters as they grapple with these challenges provides good authentic assessment opportunities for the teacher.

It's true that, without good self-esteem, students will not learn. It's also true that genuine self-esteem comes from honest achievement. By seeing to it that students struggle a bit with low-risk math problems such as those contained in this book (*low-risk* meaning that grades are not affected by them), teachers offer students the chance to feel authentic accomplishment. Students who bring that sort of feeling to class are much more likely to be enthusiastic and cooperative, and less likely to become disruptive. The need for attention will still exist, but it will tend to be satisfied by the student's active involvement in lessons and accompanying activities.

ABOUT "COUNTERINTUITIVE" PROBLEMS

All of the problems and activities in this book are designed to be of high-interest for most middle school students. Many are specifically "counterintuitive" problems. That means their answers and solutions may not "seem right" to students (or adults) at first. They are not necessarily trick questions; rather, the answers simply appear to fly in the face of common sense, *before discussion*, that is. Therein lies one of their benefits to math teachers and students; these problems promote confusion and debate, followed by understanding.

As an illustration, I point teachers to perhaps the most quintessential counterintuitive problem in this volume, "Birthday Match," Problem #21. That it takes only 23 persons in a room to assure that there is a 50/50 chance of having two matching birthdays seems downright absurd. Even after its accuracy has been verified, one may have difficulty ignoring that little voice that keeps whispering, "But that can't be right!" over and over again. That dissonance is part of the fun inherent in these types of problems and activities. Adolescents tend to be passionate about defending that which they see as right, and that is why teachers will find the problems and activities in this book so helpful. Even students who are usually reticent to participate will find themselves being drawn out by these challenges. Motivated students will not only pay attention, but will think and reflect, be anxious to contribute, and perhaps most importantly, will want to continue to contribute even after being "wrong."

The types of problems and activities presented in this book should be shared with students daily, if possible. They should not be seen as departures from "real work." Teachers should never close out the exploration of these activities with a statement such as "Now let's get back to work."

CHARACTERISTICS OF ADOLESCENTS

It has already been mentioned that adolescents seem to enjoy a good verbal fight—particularly when they have been given implicit permission to fight, and when the opponent is an adult. This book provides good opportunities for teachers to encourage that tendency, while defining the characteristics of a *fair* fight for the students. What are other characteristics of the early teen years that teachers need to keep in mind while using the problems in this book?

(*1*) Adolescents need attention, and they will generally do whatever is necessary to garner it. Every obnoxious act is a plea for attention. One way or another, the child will get that attention!

(*2*) The need for status is another powerful force in adolescents' lives. Status can be defined as the need to be significant; to somehow matter. For many middle-grade students, it is the driving force in their lives.

Teacher Question Box

When the need for status is overwhelming, what might extremes look like, for students or for adults?

(*3*) They can be talkative—often without thinking.

(*4*) They may blossom (or regress) quite unexpectedly. This is true in the academic, social, and emotional areas.

(*5*) Their behavior may tend toward defensiveness, moodiness, sullenness, and the downright odd.

(*6*) Many adolescents are sensitive to criticism, but can "dish it out" with no problem.

(*7*) Adolescents' moods are inconsistent and unpredictable. Like all people, they are complex creatures. They may be morning people, or night owls. They may be visual or auditory learners. They may, or may not, need movement in order to learn. They possess strength in several of the multiple intelligences, and underdeveloped capacities in others. Their preferences at any given moment may depend upon moods, physical health, eating and sleeping patterns, and the status of their home life and peer relationships.

What adults generally refer to as obnoxiousness and misbehavior are in reality perfectly normal (and rather thin-veneered) developmental characteristics of the pre- and early teen years. At the same time, those characteristics are also reasonably labeled as unacceptable by adults (although good teachers must appear to remain completely unimpressed with even strong student misbehavior). In any case, student docility should never be a teacher's goal.

Teacher Question Box

Is the display of sincerely shocked, insulted, or offended reactions from adults ever appropriate? If so, when?

Teachers should expect odd, defensive, attention-getting behaviors from adolescents. Shocked, insulted, offended reactions from adults are nearly always inappropriate. In fact, those types of reactions may reinforce the attention-getting behavior. Of the shocked adult it can be asked, "What were you expecting?" Abnormal behavior is normal behavior for the adolescent.

Teacher Question Box

How can student misbehavior be simultaneously normal and unacceptable? Can you think of an example that illustrates this possibility?

THE EFFECTIVE TEACHER

Use of low-risk, high-interest problems and activities, such as those found in this book, is not in and of itself enough to assure success for all students in mathematics classrooms. Most informed, caring, modern teachers know that some teaching techniques work some of the time, and that some don't; hence, the *art* of teaching. They know as well that the tone set by the teacher is as important as the use of effective teaching methods. Here are reminders for teachers who are attempting to make their classrooms as learner-friendly as possible:

Acceptance

Being fair and non-negative is not enough at the middle-grade level. Teachers need to be actively doing those things that help to meet youngsters' developmental needs. Just *expecting* students to be "good" because a teacher is essentially fair is naive.

The single most important factor in student motivation in the middle grades may be this: to what extent does the teacher offer unconditional acceptance and safety? How does a teacher confirm these things with students? By stating "I accept you unconditionally"? Probably not. Instead, that confirmation is likely made (or dispelled) when a child is at his or her worst. This is not the same thing as "nice-ing" kids to death or being a doormat. The keys here are setting reasonable, age-appropriate limits, and the separation of the person from the behavior.

If a student doesn't have the need for acceptance (and for some power control and some recognition) met somehow, he or she will likely not cooperate with the teacher. The size of any given punishment for misbehavior will not matter; "If you're thirsty enough, you'll drink muddy water."

Teacher Question Box

Can you think of an example from middle school of an incident that required teacher acceptance of the person, but not the behavior?

Trust

Building trust with students—it's an essential element in the promotion of student risk-taking. Here are some teacher behaviors which will, and some which will not, promote student trust:

Will Promote Trust

- promising and keeping confidentiality
- making, then keeping, promises
- apologizing when appropriate
- listening to students, using eye-contact (more on this below), focusing, reiterating, drawing out, etc.
- complimenting students, especially when this is not expected
- allowing and encouraging risk-taking
- remaining approachable
- being completely accepting of students as individuals

Will Break Down Trust

- sharing confidences (including sharing them with colleagues who do not have a need to know)
- using sarcasm (even if "richly deserved"—see discussion below)

- use of personal or group insults (The degree of pointedness of a teacher comment should be judged by the recipient, not the sender.)
- holding grudges; reminding students of past shortcomings
- eye-rolling and other negative body English
- exuding a "don't bother me" attitude

A brief note regarding sarcasm: Teachers should avoid "burning" students with sarcasm, especially when students have set themselves up nicely for zingers. Having said that, it must be noted that there can be occasions when a teacher might allow some good-natured ribbing to occur (teacher-to-teacher, student-to-teacher, student-to-student) in a middle-grade classroom. Caution must be exercised, however, and the teacher must first make a sober judgment regarding the appropriateness of the comments. Properly used, those kinds of exchanges *can* be signs of a healthy classroom.

Teacher Question Box

Can you describe what an appropriate use of sarcasm in the classroom might look like?

The effective, learner-friendly teacher knows the value of "catchin' 'em being good," as opposed to waiting for students to break rules so that they might be disciplined. One of the most influential things a teacher can do to front-load student success is to make surprise, strictly positive telephone calls to the parents of difficult students. This can be life-changing for some youngsters. (Surely every child has some characteristic which can be the subject of such a telephone call!) We do know that a limited number of individuals are allergic to praise, but the risk here is small.

Eye Contact and Touch

Good teachers never forget the power of eye-contact (This is not a reference to the "teacher glare"!) and an appropriate touch. The handshake and the high five, fortunately, have not yet come under attack in America. (The handshake can take the place of a hug in a litigious society.)

Teacher Question Box

What are other safe and legitimate ways for teachers and students to make physical contact?

Motivators

Here are some teacher behaviors that are likely to increase student motivation:

- modeling interest at all times
- being consistently enthusiastic
- inducing curiosity
- communicating expectations
- providing meaningful feedback to students
- structuring activities as learning experiences, not just as preparation for tests
- adapting to student interests
- providing some student choice
- providing higher-level thinking and divergent thinking questioning opportunities
- providing opportunities for frequent interaction with peers

To connect with middle-grade students, the teacher should make experiences novel as often as possible (but always with purpose!). Use of variety and gimmicks should be part of daily classroom life.

To further enhance student motivation, the teacher should see to it that teaching is as personal as possible. Adolescents enjoy it when the teacher makes references to elements of their personal lives—carefully done, of course. This can include asking students about extracurricular activities, and it can mean weaving references to students' lives into lessons. The teacher can insert those references into daily math lesson examples and illustrations. ("When Joan ran her mile race in last Tuesday's track meet, what was her average time per lap if. . . .") This approach caters to the adolescent need for attention, and it does it in an appropriate manner. Teachers of middle schoolers should also link lessons and activities to popular culture and modern students' interests, without attempting to be the students' good buddy (and without necessarily endorsing all themes of popular culture and attendant fads)!

Teacher Question Box

Should teachers attempt to act younger than they are for the sake of relating to students? Should they demonstrate their knowledge of current pop culture?

Another good motivator teachers may want to consider in math classrooms is this: When students learn something new as a group, the teacher should permanently post in the room a summary of what it is they have learned. Better yet, the teacher might assign the task of summarizing what has been learned to a student or to small groups of students. Their job might be to produce an attractive but concise poster or display based upon the lesson or unit just completed.

Teacher Question Box

Can losing ever motivate?

Level of Concern

The students' daily level of concern should be appropriately set. The teacher should afford students the opportunity to struggle, even to fail, and to learn from the experience. If the teacher provides no reason at all for students ever to be anxious, some may have no motivation to participate. Likewise, setting a level of concern too high can result in student frustration and shutdown, and inaccurate assessment results. (Poor scores on tests can be the result of panic.) Providing *some* concern for students is the teacher's goal. This underscores the idea that while the teacher is on the students' side, he cannot be their good buddy. (We certainly don't want to put boulders in their paths, but do we want to give youngsters smooth-as-glass highways? No potholes, ever? Shall we attempt to continually structure a child's world so that he will never encounter any difficulties?)

Teacher Question Box

Why might assessment results be inaccurate when the student level of concern is too high?

Informing Students

Effective teachers let students know what it is that they are going to learn, *in advance*. Good teachers don't use the statement "Here's what we are going to cover today." Instead, they use statements such as "Here's what we are going to learn about today." Better yet, after a review of where the students have been of late, good instructors ask "Based on what we've learned so far, what do you think we're going to learn about now? How do think we're going to extend our education?"

Questioning Technique

Use of good questioning technique may assist in setting that proper level of concern. Use of wait time is essential when questioning. Refusal to call on the first student who raises his or her hand is wise. Announcing that "I'm going to call on whomever I want after I ask this question—be ready with an answer," is effective in keeping students actively participating. The teacher should rarely repeat students' answers or comments aloud—others should be made to listen to the speaker. If some claim that they did not hear, the teacher should ask the speaker to repeat the comment. The smart instructor will not always label a student's response as right or wrong. Instead, she will ask the student to elaborate or to provide rationale, and will often refer to other class members for their reactions to the speaker's comments.

The teacher should turn announcements and statements into questions whenever possible. This will help to promote active learner participation. For example, instead of stating that "Our quiz on factoring will be given this Friday," the teacher may want to try "So, when do you think I'll be quizzing you regarding our study of factoring?" This might be followed by "Why should I quiz you then?" This approach actually helps to cement previous student learning.

An effective extension of this is the technique of letting students say what needs to be said, as opposed to the teacher always saying it. For example, instead of "Today you'll learn how to factor trinomials wherein the squared term has a coefficient other than an invisible one," the teacher should try to pull off an exchange like the one that follows (admittedly a somewhat simplified version, but it gives a flavor of what direction the teacher ought to be going in).

> **Teacher:** "What have we been learning about for the last few days? Please be ready with an answer in 10 seconds—I'll be calling on whomever I feel like calling on." (Wait time here, of course.)
> **Donna:** "How to factor trinomials."
> **Teacher:** "Right. But have we been factoring just any old trinomials?" (Teacher writes on the board or overhead screen several familiar trinomials, all with one (1) as the first coefficient.)
> **Class:** (no response)
> **Teacher:** (underlining the invisible ones that precede each squared term) "What don't we see here, and here, and here? These are simple, wimpy trinomials. Why?"
> **Kathy:** "The squared terms have no coefficients."

Teacher: "Good!" (Teacher pauses, hoping that someone will correct Kathy's small error—the squared terms do have coefficients, they're just invisible ones.)

Teacher: "Actually, the coefficients are all ones, but we knew what Kathy meant."

Teacher: (again underlining the invisible ones as a hint) "So, what new skill do you think we'll be learning about today?"

Sam: "How to factor trinomials with coefficients that aren't one?"

Teacher: "Bravo!"

Note the use of varied terms of praise. "Good," "Way to go," "Yes, indeed," "How about that?," "Wonderful," "Splendid," "Perfecto," "I am impressed," etc.

What are the advantages of this admittedly longer, time-consuming approach? It brings in and reviews prior learning, and it provides an anticipatory set for students. (This is a modified Socratic approach; that is, the teacher assumes, in a sense, that students already have most of the knowledge with them, and that the teacher's job is to bring that knowledge out.)

Here are some other important questioning techniques:

(1) The teacher should use variations of the following statement often: "I'm going to call on anyone I like, whether or not your hand is up, so please be ready with an answer. It's OK if your answer isn't perfectly right, but please have an answer ready."

(2) It is wise to remember to use wait time when asking questions. Those ten seconds can be a bit uncomfortable for both teacher and student, but that's alright. Wait time gives everyone a chance to think. When paired with tip *(1)* above, it can be dynamite.

(3) The teacher should let frequent hand-raisers know that they will only get to answer or contribute two or three times per lesson, so they should save their comments for their best answers.

(4) Have all the students' names written on sticks, ready to be randomly drawn. This provides an effective method of choosing students to answer questions or to contribute.

(5) The "Bonus Question of the Day" technique should be considered. Here's how it works: Each day, the teacher designates, in advance, the "Bonus Question of the Day." This can be any question, trivial or profound, that is likely to be asked by the teacher over the course of the day. When the teacher asks the Bonus Question—remembering that the students don't know in advance which question it will be

that day—the teacher should make a big to-do over whoever answers it. A small extra privilege, perhaps? A stock certificate? (But *never* an improvement in the student's mathematics grade!) If the question is answered correctly to the teacher's satisfaction, the award might be doubled. (It's best to make the Bonus Question black and white, that is, one with an unambiguous answer.)

(6) The effective teacher will ban this sentence from his/her vocabulary: "Are there any questions?" It is surely one of the most useless queries a teacher can ever make. It is not checking for understanding—what middle school child is likely to ask a question moments before dismissal time? Here is much better wording: "I'm going to make some statements regarding today's lesson. If I call on you, give me thumbs up or thumbs down to indicate whether or not you think the statement is true. Here I go. The volume of a room refers to how much carpeting it would take to cover the floor. True or false?"

Here's another example of a better approach when checking for understanding: "I'm going to purposely make an error in my calculations on the chalkboard. Try to spot it, because I'm going to call on several of you at random and ask for your thoughts." An extension of this technique, designed to keep everyone mentally involved for as long as possible, is for the teacher to call on other individual students to agree or disagree with the previous student's comments.

(7) Good questioning technique may have teachers frequently using phrases such as these:
- "Can you explain . . . ?"
- "Why do you think that . . . ?"
- "Tell us more about . . ."
- "Is it fair to conclude that . . . ?"
- "Can someone else paraphrase . . . ?"
- "Then, can you make a prediction . . . ?"
- "What's your bottom line conclusion?"

Related to questioning technique is *direction-giving*, and the effective teacher knows that time is a great motivator in this area. Which is better: "Please get out your paper and pencil in the next 15 seconds," or "Get out your pencil and paper"? Of course, the teacher must be ready with a follow-through if paper and pencil are not out in 15 seconds! What might be appropriate in that case? A frowny face? Perhaps—it *can* work. Withholding a minor privilege? Yes. Assigning additional homework? No—never. Teachers need to break this unfortunate school tradition. This practice flies in the face of teachers trying to convince youngsters that learning can be challenging but immensely satisfying.

Using the method known as "ticket out" can help to set that proper student level of concern. This technique has the teacher stationed at the exit door as students are leaving, with their "ticket out" of the door being the requirement of showing the teacher that a given task has been accomplished. For instance, the teacher may state that the ticket out is "five of today's math exercises done" or "at least half of this morning's writing assignment completed." Naturally, students should be told in advance what the ticket out will be. Having a little notebook for recording purposes at the door may be useful.

Teachers have the right to define "ticket out" differently for different students. "Every problem must be done" is not generally a reasonable ticket out. As educator Dr. Richard LaVoie has said, "Fairness is not treating everyone the same, fairness is giving everyone what they need."

What might, say, three "checks" in that ticket out notebook indicate? The loss of a classroom privilege is one idea. (Repeat warning! The lack of a student ticket out should *never* result in a lowered mathematics grade.) By the way, the loss of privileges naturally implies the existence of those privileges—as it should. The teacher should never give the desperate child a chance to say "What have I got to lose?"

Note: The teacher may also want to try teaching a lesson silently, using only the chalkboard or overhead projector, hand and facial signals, and "body English." The teacher may find that students will pay closer attention than usual, and that they will take some pride in being able to learn the lesson without the teacher saying a word. (This may also serve to point out that most teachers do much more talking than is necessary in the classroom.)

Movement

The effective teacher avoids group instruction that lasts more than 15 or 20 minutes, unless it is a lesson of the highest interest to middle-grade students. If group instruction becomes group *activity*, much more time can be spent in this teaching mode. If that activity can involve controlled large-scale movement on the part of youngsters, they are likely to remain attentive and cooperative. Here are some examples of how movement can easily be incorporated into math lessons:

(*1*) Frequently conduct instant student-to-student surveys. For instance, during the middle of a lesson on banking and mortgages, the instructor can tell students "When I give you the signal, get up and ask three different people to define 'mortgage' for you in twenty words or less. It's OK if someone says 'I'm not sure.' Jot down what you think is the best definition, and be prepared to share it with the rest of the class. You'll have two minutes for this—I'll keep track of the time."

Another "instant survey" might look like this: Have students ask exactly four other students in the class how many siblings they have, then have them make a guess as to the typical number of siblings for the entire class (the mode). This particular survey sets up a discussion of *sampling*. ("Is talking to four classmates out of 26 enough to make a good guess? If your four had all said 'none,' why did you still guess that 'two' was the right answer?")

(*2*) Ask students questions that require them to move about in order to find the answer. For instance, the teacher can have students measure the length and width of the classroom by pacing them off (with their *own* paces, of course). Students can then report their findings; the teacher or a designated student can put all results on the chalkboard; and students can be asked to discuss (or write about) why responses varied.

The teacher should build in natural movement opportunities for highly kinesthetic youngsters, or youngsters diagnosed with Attention Deficit Disorder. For example, the teacher may wish to make Jimmy the official classroom messenger person. Once per hour, she sees to it that Jimmy delivers an "important" sealed message to the office or to another classroom. The contents of the message, of course, may or may not be substantive. Another accommodation for students who truly need to move about is to allow them to stand at the back of the room during group instruction or other times (with the agreement that the students will focus their attention on the teacher). Simple cardboard podiums can also be provided at the rear of the room. Students can listen from that vantage point, and even write there. Providing cotton swabs for incessant pencil-tappers is another idea to try with kinesthetic learners. Use of these items will be novel at first, but the novelty will soon wear off, and only those students who "need" to tap will bother with them.

Teacher Question Box

What are the pros and cons of allowing standing and movement? If you see movement as beneficial, can you identify other ways of making movement an option for students?

Dead Time

Avoiding class "dead time" is important. This is most likely to happen at the beginning of class, the end of class, before and after transition times, and when small groups or individuals finish activities. While it's certainly true that good teachers need a few moments here and there (in addition to lunch and preparation times) to catch their breath, this is not the issue here. The problem arises when a short pause stretches into five, then ten minutes of unstructured time. As most educators know after their first year of teaching, "dead time" is when discipline problems can occur.

How to avoid dead time? One way is for the teacher to regularly make use of warm-ups, cool-downs, and sponge activities—all with a purpose. A key criterion for determining the appropriateness of these activities is this: Can a student accomplish the task with little or no teacher help? Furthermore, they should be fun, engaging, and generally concise tasks. The classic "board work" concept is fine, if the tasks are meaningful. The teacher can also make available numerous small sheets of paper with quick games on them—tic-tac-toe and "brussel sprouts," for instance. Providing shoe boxes full of hands-on games (one-on-one or individual) for student use is another good idea (see the discussion regarding this idea in Appendix C).

Incidentally, dead time's cousin, "free time," is not a good thing in middle-grade classes. Offering free time as a reward, in particular, can send the message that learning is a thing to be avoided.

Closure

Avoiding lesson "closure by default" is crucial. An example of closure by default is this: The school bell rings, students need to run to their next class, the teacher needs a moment to prepare for the next group, and that's that. Instead of that scenario, closure should be planned, and it should be more than just the teacher summarizing the lesson. For closure to be student-centered, students need to be asked questions that will help cement their understanding of the lesson and which will provide the teacher with some assessment of the general level of student understanding. Closure can *include* a summary, but closure and summary are not the same thing. Actively involving students in closure will likely make it more effective. ("I'm now going to ask whomever I feel like calling on to tell me—in your opinion—which is the most important step when computing discount, and if you're called on, plan to give us a rationale for your answer.") If the teacher simply provides the summary, there are no guarantees that students are learning, or even paying attention. Having students directly involved in closure helps to maintain that appropriate level of concern. (By the way, asking students to provide a rationale for answers and comments ought to be a regular and predictable feature of math classrooms.)

Patience

Real estate agents say that the three most important factors in determining the value of a house are location, location, and location. Wise educators know that the three most important factors in determining teacher success in the middle-grade classroom are patience, patience, and patience. This implies the maintenance of a broad perspective and a durable sense of humor. (Students may groan at a teacher's corny jokes, but they actually appreciate the sentiment.) Many a potential classroom crisis can be averted with a well-timed teacher-provided joke. (Not every student rule infraction needs to be dealt with on the spot.) More on this in Appendix C.

Forgetting

Educators should keep in mind that forgetting is natural. We all forget much of what we have "learned," and nearly all of what we see and hear. Saying to students, "But we covered that!" is not a legitimate instructor statement. "Covering" curricular material, even covering it well, doesn't guarantee that students have learned it for the long run.

Teacher Question Box

What can be a dangerous pitfall for teachers in late May or in June regarding "covering" the curriculum?

"Writing Students Off"

Caring teachers never will write off a thirteen-year-old, regardless of how objectionable his or her behavior may seem, or how hopeless future scenarios may appear. As mentioned earlier, adolescents may blossom unexpectedly. Current behavior is not necessarily a reliable predictor of long-range future behavior.

Other effective teacher behaviors include:

- always grading, or in some way responding to all student assignments, homework in particular
- using calculators regularly—research is clear that their use can increase student achievement and improve attitudes
- not letting the *perfect* stand in the way of the *good* (Effective teachers will not hand-wring or hesitate simply because a lesson has not gone perfectly, or because they have missed the bulls-eye regarding a piece of teaching/learning research.)

A good test of a teacher's overall effectiveness is this: If a visitor stopped in a teacher's classroom unexpectedly, would the visitor easily detect the following?

(1) Strong vision on the part of the teacher, and enthusiasm on the part of the students?

(2) A rigorous, integrated, interdisciplinary approach?

(3) A theme of mutual respect?

The ultimate test for teacher effectiveness may be this: Can the teacher be accused of being an advocate for children?

CONCLUSION

Good middle-grade mathematics teaching will actively engage students, and will help them to acquire a *sense* about numbers. It will also provide for spirited but guided classroom discussions, and for writing opportunities, centered on the theme of problem solving. The problems, puzzles, investigations and demonstrations that follow should provide both teachers and students with many hours of enjoyable and educational challenges.

Note: Below each problem or activity "curricular areas" are given as an aid to teachers. It can be assumed that "problem solving" is one of the curricular areas that every problem or activity touches upon.

REFERENCES

Epstein, J. L. and MacIver, D. J. (1992). *Opportunities to Learn: Effects on Eighth Graders of Curriculum Offerings and Instructional Approaches.* Baltimore, MD: Center for Research on Effective Schooling for Disadvantaged Students.

Kirsner, S. A. and Bethell, S. (1992). *Creating a Flexible and Responsible Learning Environment.* Report # TL-RR-92-7. East Lansing, MI: National Center for Research on Teacher Learning.

Rasch, K. (1994). "The imperative for quality middle school mathematics curriculum and instruction. *Midpoints,* 4(2):1–16.

CHAPTER 2

Counterintuitive Problems

#1: THE BUSY LAWYERS

Curriculum Areas: Fractions, Decimals, and Percent

"MR. BARR AND Ms. Writ are two highly competitive injury lawsuit lawyers. How competitive are they? So competitive that, twice each year (in June and December), they compare their trial win/lose records, with the 'winner' treating the other lawyer to dinner. (Mr. Barr and Ms. Writ each 'take on' a *lot* of cases!)

For the first half of a recent year, Mr. Barr's 'winning percentage' (the number of cases he won divided by the total number of cases he handled) was better than Ms. Writ's winning percentage. For the second half of the same year (looking only at cases from the middle of the year), Mr. Barr again had a better winning percentage than Ms. Writ.

Question: Is it somehow possible that Ms. Writ had a better winning percentage than Mr. Barr *for that entire year?* If so, how? If not, why not?"

Answer: Yes, Ms. Writ could have had a better all-year winning percentage than Mr. Barr, in spite of him having the better first half and the better second half. The explanation is found below.

Intuition screams loudly that Ms. Writ could not possibly have had a better year-long winning percentage than Mr. Barr. He "beat" her in the first half, he "beat" her in the second half; case closed. Although that is a good "opening argument" from students, the teacher should challenge them to push their thinking beyond the obvious. If students do not bring up the point on their own, the teacher should ask them whether or not the *number* of cases each lawyer handled is relevant. This may or may not spark some students to examine the situation more closely.

The following is a detailed description of one possible scenario in which Mr. Barr is the "winner" for the individual halves of the year,

21

and Ms. Writ, the year-long "champion." Note that the winning percentage is given in baseball-style statistics, that is, as a decimal shown to the thousandths place.

First half of the year
Barr: 60 wins out of 100 cases tried; winning percentage: .600
Writ: 29 wins out of 50 cases tried; winning percentage: .580

Second half of the year
Barr: 40 wins out of 50 cases tried; winning percentage: .800
Writ: 79 wins out of 100 cases tried; winning percentage: .790

Entire year
Barr: 100 wins out of 150 cases tried; winning percentage: .667
Writ: 108 wins out of 150 cases tried; winning percentage: .720

Note: If one simply takes the mean of the first-half and second-half winning percentages of either Mr. Barr or Ms. Writ, that average is not the same as the actual year-long winning percentage of either lawyer.

An everyday school example may help to clarify the issue for students: Jean is given only one mathematics test during the first semester of school, and on it she earns a score of 100%. During the second semester, twelve tests are administered, and Jean consistently earns a score of 80% on each one of them. Is it fair to take the first semester score of 100% and average it with the consistent second semester score of 80% for a yearly average of 90%? Jean might wish it so, but that's bad mathematics!

Students should walk away from this activity with the understanding that *one cannot average averages* (unless all of the averages under consideration have the same base, or "denominator").

#2: WHY ISN'T ALL THE MONEY GONE?

Curriculum Areas: Consumer Mathematics, Percent, Interest, Compound Interest, and Banking

"The net worth of Leo's Jam and Jelly Company at the start of a particular year was $32,000. Due to declining sales, Leo's company began losing 10% of its worth each and every month. At the end of the year (twelve months later), the company was still worth $9,037.74. How can this be? Wouldn't losing ten % per month for just *ten* consecutive months, let alone twelve, leave Leo with a worth of $0?"

A discussion of this problem with middle schoolers will help to clarify the truth that 10% (for instance) of a dwindling amount is an ever-dwindling amount. Ten percent of something remains a consistent size only if the original amount under consideration does not vary.

The teacher's goal here is to lead students to state, in their own words, the phenomenon noted above. A praiseworthy observation would be a student comment such as, "Well, when you chop 10% off of what you have, you have something smaller, and 10% of *it* will be smaller, too."

Extension Activity

If students seem to grasp this problem's underlying concept, the teacher may want to use it as a lead-in to a look at the magic of compound interest. Here's a good introductory problem:

"If you deposit $1,000 in the bank, and the bank adds 1% to your savings each month, which of these amounts represents your balance after 10 months? (Let's assume that you do not add or withdraw any money for those ten months.)

(*a*) $1,104.62   (*b*) $1,100.00"

Naturally, if any students choose (*a*), the teacher should ask for a rationale. Some students will choose (*a*) without knowing why; some will be able to offer a rationale. If the rationale is sound, the teacher should then reinforce the concept of compounding.

Others in the classroom may agree that the answer is $1,104.62, but may insist that the difference of $4.62 is minuscule, and that compounding, therefore, is of insignificant importance. The following activity provides a way to counter that thinking.

The teacher should "give" students, say, $5,000, and ask them to add interest to this amount two different ways:

Method #1: Add 10% of $5,000 ($500, of course) to $5,000, ten times. (Students will now have $10,000.)

Method #2: Add 10% of $5,000 to $5,000 one time. Now take 10% of that new balance, and add it to the balance. Keep taking 10% *of the latest balance* and adding it *to the latest balance*; do this a total of ten times. (Students should use a calculator for this exercise—the balance will be $12, 968.71.)

Most students will agree that the difference between the two final balances is substantial.

Here are two other activities which will help to reinforce students' understanding of the potential power of compounding:

(*1*) The instructor should "give" students $100, then allow that money to grow by 50% over and over again by multiplying the balance by 1.5 (representing 100% + 50%) on the calculator. The teacher should then ask students to predict how many steps it will take to increase that $100 to $1,000,000. Few will predict an answer as low as 23 steps, but that is correct. ("Steps" in this problem represent compounding periods. Naturally, 50% is unlikely to be the actual monthly, or even annual, rate of interest. More realistic monthly and annual rates of interest should next be discussed with students.)

(*2*) The instructor should let students experiment on their calculators to discover what percentage rate of interest will allow $25 to become $1,000 in, say, 20 steps. (Remember if a student decides to try using, for example, a rate of 12%, she will need to use 1.12 as the multiplier.) One way for the teacher to present this type of problem to the students is this: "How can I make $25 into $1,000 in 20 easy steps?"

The teacher will undoubtedly be able to think of many variations on the above problems.

The next problem should be presented *after* students have wrestled with "Why Isn't All the Money Gone?" Their experience should prove valuable; indeed, the teacher should ask students to reference this problem as they look at finding out "What's the Cost?", the next problem in this section.

Teachers' Review of Percent Shortcuts

Ten percent of any amount is found by moving the decimal point (even if "invisible") one place to the left. Examples:

- 10% of 600 is 60. (The decimal points are "invisible" in this example.)
- 10% of 2,407 is 240.7.
- 5% of any amount can be found by taking 10% of the amount,

then *half* of that 10%. An example: 5% of 460 is 23 (*half* of 46).
- Using the simple methods for finding 5% and 10% of varying amounts allows one to easily compute many other common percentages, such as 15%, 30%, etc. One example: 15% of 920 is 138 (92 plus *half* of 92.)

#3: WHAT'S THE COST?

Curriculum Areas: Consumer Mathematics, Percent, Discount, and Markup

"A pair of slightly out-of-style bluejeans displayed in Mr. Riggers' clothing store had a price tag of $50, and no one seemed interested in buying them. The store manager decided to reduce the price by 10% in hopes that they would now sell. They didn't. Mr. Riggers thought, 'If it won't sell at 10% off, I might as well raise the price again. Maybe I'll get lucky.' So, he *raised* the price by 10%. What was the cost of the bluejeans after these two price adjustments (ignoring tax)?"

Most students will quickly give an answer of $50. After all, if the price went down by 10%, and then went up by 10%, shouldn't the final cost be the original $50?

No. Here's why:

The price did *not* go up by the same amount that it went down (even though both were 10%). When the price first went down by 10%, the sale price became $45, ($50–$5), as most students will be able to state. When the price went back up, it went up by 10% of $45, not 10% of $50. Thus, the price tag finally reads $49.50.

#4: THE PEN AND PENCIL SET

Curriculum Areas: Algebra and Logic

"Together, a pen and a pencil cost $1.10. The pen costs one dollar more than the pencil. What is the cost of each?"

"The pencil costs a dime, and the pen costs a dollar." That, nearly without fail, is the first response from students when presented with the Pen and Pencil Set problem. It is, of course, wrong. The difference between ten cents and one dollar is not one dollar.

The teacher should avoid the temptation to ask students if the difference between a dime and a dollar is the one dollar called for in the original problem. Students will note that themselves in short order. Brainstorming, centering on the use of trial and error (a fine problem-solving strategy), will likely provide students with the correct answer within a minute.

(The solution? The pencil costs 5 cents; the pen, $1.05.)

After solving the Pen and Pencil Set problem, students may be better equipped to solve another similar problem:

"A watch and battery together cost $18.00 (ignore tax).The battery cost $10 less than the watch. How much did each cost?

Answer: The watch cost $14, the battery, $4 (not $10 and $8, as some students will offer).

If algebra is to be used (and use of algebra would be a typical middle-grade math classroom approach to finding the solution), an equation that works is

$$x + (x - 10) = 18, \text{ giving } x = 14.$$

It should be kept in mind that while algebra is an excellent method for solving many problems, its effectiveness can often be enhanced by allowing students to explore and use other intuitive methods as well.

One more problem in the same vein as the two presented above is "What Does the Clock Cost?" which follows.

#5: WHAT DOES THE CLOCK COST?

Curriculum Areas: Time and Algebra

"A clock costs $10, plus half of its cost. What is the cost of the clock?"

Fifteen dollars is often, and incorrectly, offered as the answer. However, if the two amounts known as "ten dollars" and "half of the cost" total the entire cost, doesn't that mean that the "ten dollars" is the other half of the cost? (Yes, it does.) The clock costs, therefore, $20. (As is often the case in these sorts of problems, we are ignoring the issue of tax. If a student chooses not to ignore that issue, this will provide a fine opportunity to discuss how injecting tax into the problem might affect the solution.)

#6: THE DITCH DIGGERS

Curriculum Areas: Algebra, Patterns, and Sequences

"If two ditch diggers can dig a total of two ditches in two hours, how long will it take four diggers to dig a total of four ditches? (Let's assume that all ditch diggers work at the same pace.)"

Nearly without fail, learners will provide an answer of four hours. The correct answer is two hours.

Let's look at the first sentence. It tells us that one ditch digger takes two hours to dig a ditch. That becomes our "base unit of work": 1 digger, 1 ditch, 2 hours. If there are four diggers, each digger still needs two hours to complete "his" ditch, so the four ditches still take two hours to be dug.

The two-two-two pattern of the original scenario may bring students to suggest that four-four-four should be the pattern for the second scenario.

The teacher may want to suggest that four students actually act out the scenario. A fifth student can act as a timekeeper, letting the student diggers know when the "hour" is over. "Acting it out" is an excellent problem-solving strategy (see Appendix A).

#7: THE RAFFLE

Curriculum Areas: Probability and Statistics

"In order to raise money for a new outdoor school sign, the Principal at Acme Middle School decided to hold a raffle. Exactly 1,000 tickets would be sold for one dollar each. One winning ticket would be chosen, the prize being free school lunches for a year. The first 600 tickets were sold to 600 separate individuals, and the remaining 400 were all sold to a Mrs. G. Bernhoot."

Question: Who will likely win the raffle?

We can't say for sure who will win, but it probably *won't* be Mrs. Bernhoot! This statement will be met with some objection on the part of students, mostly due to the observation that Mrs. Bernhoot is *400 times as likely* as anyone else to win. This observation is a correct one, but she will probably still not win. Here's why:

The sold tickets can be seen as forming two blocks, one of 400, another of 600. When those 1,000 tickets are thoroughly mixed together in a drum, neither the drum nor the ticket-drawer care, so to speak, that Mrs. Bernhoot has purchased so many. The fact is, 400 tickets are hers, and 600 *aren't*! It is likely (although by no means certain) that the winning ticket will come from the block of 600, not the block of 400 (Mrs. Bernhoot's tickets).

What makes this problem so counterintuitive is the coexistence of the following true but apparently contradictory statements:

(*1*) Mrs. Bernhoot is 400 times as likely as anyone else to win.

(*2*) Someone else (whose ticket is among the other six hundred) is likely to win.

The reconciliation of the two statements is as follows: Mrs. Bernhoot is indeed 400 times more likely than any other *individual* ticket holder to win the raffle, but she is not 400 times more likely than all of the other 600 ticket holders to win.

A good follow-up to this problem would be a discussion of the mathematical wisdom of playing state and regional lotteries. Perhaps teachers or students have read of individuals who have spent their life savings buying thousands of tickets for one day's lottery drawing, their thinking being that, if they buy, say, 5,000 tickets, and no one else has bought more than, say, 20, for that drawing, they are sure to win. After all, 5,000 is many times more than 20! Like Mrs. Bernhoot, however, that individual will probably not win. While he may have purchased 5,000 tickets (far more than anyone else), total sales might equal 1,000,000 tickets, nearly assuring that someone else will win.

Some lottery advertisements use a slogan such as, *"Someone's gonna win!"* The appropriate follow up line should be, "and it probably won't be you!"

#8: THAT'S NOT ENOUGH INFORMATION! PART I

Curriculum Areas: Products, Factors, and Sums

This wonderful brain teaser is not for everyone. It will frustrate students who are not ready to handle the math concepts involved. But for those who can, it provides a delightful opportunity to wrestle with a clever and unusual challenge.

"A census taker rings the doorbell of a home. A man answers. The census taker asks, 'Can you please tell me how many children live in this house?' The man answers 'Three.' Next, the census taker asks,

'And what are their ages?' the homeowner, being a witty and clever man, says, 'The product of their ages is 72.'

The census taker, being pretty smart himself, knows that many different number trios can give a product of 72. He states, therefore, that 'You haven't given me enough information,' to which the homeowner responds, 'Well, my girls' ages add up to my house number.'

The census taker steps back on the porch, looks at the house number, then announces that 'I still don't have enough information to know your daughters' ages!'

Finally, the homeowner says, 'My oldest daughter prefers chocolate ice cream,' at which point the census taker says 'Thank you. I now have enough information to determine your daughters' ages.' And indeed he does.

What are the ages of the three girls?"

Middle-grade students will protest that, even with the homeowner's last statement, there is not enough information available to answer the census taker's question. They will likely insist that we need to know the house number in order to solve the puzzle. We do in fact need that information, but it is available for those who dig. (The teacher should emphasize to students that this is not a "trick" question. It can be answered with the information at hand, and there is no fancy or devious word play at work here.)

The teacher can guide students to the answer as follows: Does the statement that the girls' ages multiply to 72 narrow the possibilities? Yes it does; there are a limited number of whole number trios which give 72 as a product. With some teacher assistance, students should be able to come up with these trios:

$1 \times 1 \times 72, 1 \times 2 \times 36, 1 \times 3 \times 24, 1 \times 4 \times 18, 1 \times 6 \times 12, 1 \times 8 \times 9,$
$2 \times 2 \times 18, 2 \times 3 \times 12, 2 \times 4 \times 9, 2 \times 6 \times 6, 3 \times 3 \times 8, 3 \times 6 \times 4$

We can assume that the census taker is sharp enough to quickly come to the conclusion that one of these trios represents the girls' ages—but which trio? When the homeowner informed the census taker that the ages added up to the house number, and the census taker looked at the number, shouldn't that have provided him with enough information to answer the question? For instance, if the house number was 23, wouldn't the census taker then simply pick the trio whose number added to 23? (That would be the 1-4-18 trio, because $1 + 4 + 18 = 23$.) What could be the only reason at this point that the census taker would say, "I still don't have enough information"? The answer is this: If the house number was 14, (and *only* if it were 14), the census taker would still be stumped, because two of the trios add up to 14! How would the census taker know in that case which trio was correct (2-6-6, or 3-3-8)? A house number of 14 is the *only* house number which presents the census taker with this dilemma, so that *must* be the house number.

The census taker now knows that the girls' ages must be either two, six, and six, or three, three, and eight. But which trio? When the homeowner stated that "my *oldest* daughter prefers chocolate ice cream," the census taker knew that the three-three-eight trio was correct—there is no oldest daughter if their ages are two, six, and six! Their ages are indeed three, three, and eight.

An important point for the teacher to make here is that what sometimes appears to be minimal or downright insufficient information may be adequate after all. Jumping to early conclusions can be dangerous!

Note: Middle schoolers will enjoy debating whether or not the trios such as 1-1-72 and 1-2-36 should even have been under consideration, due to the unlikelihood of a man having daughters of those ages.

Teachers' Review of Vocabulary

Teachers should use the terms product, quotient, sum, and difference with students often. Defining them isn't enough; they need to be used. A review:

- A *product* is a quantity obtained by multiplying two or more numbers together (those numbers are called *factors*).
- A *quotient* is a quantity obtained by dividing two numbers together.
- A *sum* is a quantity obtained by adding two or more numbers together (those numbers are called addends).
- A *difference* is a quantity obtained by subtracting one number from another.

#9: THAT'S NOT ENOUGH INFORMATION! PART 2

Curriculum Areas: Geometry, and Area

When the teacher first asks students to find the areas of the figures below, most students will protest that not enough information has been given. This affords the teacher the chance to again reinforce the concept that, sometimes, minimal information may be enough information (Figures 1–4).

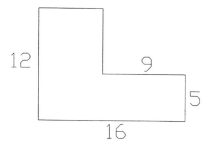

Figure 1.

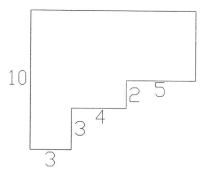

Figure 2.

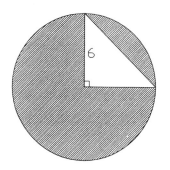

Figure 3.

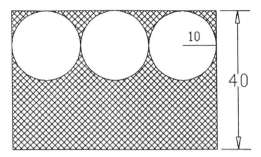

Figure 4.

The key to solving these is for students to attempt to put two or more basic bits of information together to provide additional basic bits of information. In Figure 1, for instance, the height of twelve centimeters noted on the left, paired with the partial height of five centimeters noted on the right, gives enough information to state that the missing partial height on the right must be seven centimeters. Applying the same technique on the horizontal segments completes the process of discovering the needed basic information. With that, the polygon can be "sliced" into two rectangles, easily allowing the total area to be discovered. That, and similar approaches, will work for all of the figures shown. Assume that pi = 3.14, that sides that appear to be straight are straight, and that corners that appear to be right angles are right angles. (Areas of the given figures are noted below.)

Extension Activity

The teacher should challenge students to come up with their own complicated (but clean, neat, and complete) area problems, accompanied by the minimum of information needed to solve them (as an optional assignment). If possible, students should draw their figures on large poster board or easel newsprint. Another option is for the teacher to convert student figures into overhead projector transparencies.

The areas of the four figures shown are, respectively, 129, 83, 95.04, and 1,458 square units.

#10: ROLLING THE BLOCK

Curriculum Areas: Geometry, Circles, and Pi

"Imagine that workers of an ancient civilization are attempting to roll a stone block along a series of logs. (Figure 5 illustrates the situation.) The diameter of each log is 10 inches. If the block is pushed forward, how far (in relation to the ground) will the block travel if each of the underlying logs makes one complete roll?"

This is an interesting problem for students with some knowledge of the relationship between a circle's diameter and its circumference (pi). Those students are likely to propose an answer of 31.4 inches, because that is the circumference of each log in this problem. They should be given credit for knowing and applying the circumference formula, and for demonstrating an understanding of the meaning of pi. They should also be tactfully informed that they are incorrect. Here's why:

It's true that by rolling over the tops of 10-inch diameter logs as they make one turn, the block will have moved 31.4 inches. What students will likely fail to take into account is that as they roll, the logs themselves are also moving forward. And how far forward do the logs move in one rotation? *Another* 31.4 inches! The forward progress of the block as it rolls over the logs is assisted by the forward motion of the logs themselves. Thus, the block will roll 31.4 + 31.4 inches, or 62.8 inches total.

The Rolling the Block problem provides a great way to strengthen students' understanding of circles and the ways in which a circle's parts relate to one another.

Interdisciplinary note: Some scientists believe that the lined-up logs method was the one used to transport the Stonehenge monoliths in southern England.

#11: THE RIBBON AROUND THE WORLD

Curriculum Areas: Geometry, Circles, Circumference, and Pi

"Imagine a ribbon that somehow encircles the entire world at the equator, just touching the surface all the way around. If one meter is added to the length of the ribbon, and its shape remains a circle, how high do you guess the ribbon will be off of the earth's surface?

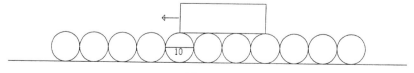

Figure 5.

(1) Approximately one-thousandth of one millimeter
(2) Approximately one millimeter
(3) Approximately one centimeter
(4) Approximately 7 inches"

Many students will choose the smallest answer, one-thousandth of one centimeter. Their thinking typically is along these lines: "That one additional meter of ribbon needs to be 'stretched' around the entire 27,000-mile circumference of the earth, so the 'raising of the ribbon' effect will be very, very slight."

Actually, choice *(4)* (7 inches) is the correct one. Here's the explanation. When the circumference of any circle is increased (as when the ribbon is lengthened by one meter), the diameter of that circle is lengthened by a little more than one-third of the size of the circumferal increase. This might be more clear if it is recognized that a circle's circumference is always a little more than triple the length of its diameter (pi, or 3.14159 . . . in particular).

If the circumferal ribbon is increased by one meter, the diameter is therefore increased by about 13 inches (about one-third of a meter). Half of the 13 inches will be evident on one side of the earth, the other half of the 13 inches will be evident on the opposite side. Thus, the ribbon, when lengthened, is approximately 7 inches off of the earth's surface.

#12: THE CARELESS CAT

Curriculum Area: Rate

Here is another classic problem which has appeared in many forms over the years:

"At precisely noon one day, a curious (and careless) cat falls into a 6-foot deep hole which has *very* slippery sides. Immediately, the cat tries to climb out, and is able to make progress at this rate: She climbs up two feet per minute, then, being very tired, slips back one foot over the next minute. When will the cat emerge from the hole?"

Many students will want to make a sketch of this situation—the teacher should, of course, encourage this. Making a sketch, after all, is a fine problem-solving strategy.

Typically, students will distill "two feet up one minute, one foot down the next" into "one half-foot every minute." The teacher should give students appropriate credit for that intelligent conversion. At a rate of one-half foot every minute, it would seem then that the cat would emerge from the six-foot hole at 12:12 P.M. That's incorrect.

What that particular problem-solving approach misses is that once the cat first reaches the top of the hole (at 12:09) we no longer have to worry about it falling back one foot over the next minute!

#13: TAKE YOUR MEDICINE

Curriculum Area: Time

"A patient is instructed to take one pill every half-hour beginning at noon. If there are 10 pills to take, when will the patient take the last pill?"

The answer is 4:30 P.M., not 5:00 P.M., as many students will first offer.

As in the Careless Cat problem, students may attempt to distill the given information incorrectly. Their thinking may be along these lines: "If there are ten pills to be taken, and each pill 'uses up' half an hour, that's five hours. Five hours from noon is 5:00 P.M." What this line of reasoning misses, of course, is that the last pill (all of them, actually) is taken at the *start* of "its" half-hour; there is no need to then consider that last half-hour.

If no students are able to state the correct answer and to give a rationale for it, the teacher may want to move students' thinking along with this question: "If the patient had to take only one pill, "beginning" at noon, would you say that it was taken at 12:30 P.M.? After all, 12:30 P.M. is one half-hour after noon, just as 5:00 P.M. is ten half-hours after noon."

#14: HIGH STAKES

Curriculum Areas: Integers and Logic

"Mr. Anderson and Mr. Bones are playing a game in which the stake is one dollar per game. When the men are done playing several games, Mr. Anderson has won two dollars, and Mr. Bones has won two games. How many games were played?"

The answer is six games. If Mr. Bones won two games, Mr. Anderson quite obviously lost them, putting him two dollars in the red. For

Mr. Anderson to come from "negative two dollars" to "positive two dollars," he must have won four games. That's a total of six games played. (The order of wins/loses here is irrelevant, and that provides a good opportunity for the teacher to reinforce the reality of the commutative property.)

#15: CUT-UP

Curriculum Area: Logic

"Jerry is handed a strip of paper an inch wide and ten inches long. The strip is marked at each inch so that it can be cut into one-inch strips. If Jerry can cut one piece per second, how long will it take him to cut the strip into ten equal pieces?"

The intuitive answer, of course, is ten seconds. (Ten pieces needed, ten cuts.) Only nine cuts are needed; the answer is therefore nine seconds. If this is unclear to students, the teacher may want to ask this question aloud: "What if you were asked to cut a two-inch strip into two pieces? Would that take one or two cuts?"

Presentation of this problem affords the teacher the opportunity to reinforce the idea that students should get into the habit of saying "Wait just a minute" to themselves when attempting to find the solutions to problems, especially when they have reason to believe that there may be a "catch" to the problem.

#16: NEED A LIFT?

Curriculum Area: Logic

"An apartment tower has 16 stories. If someone takes the elevator from the first to the sixteenth floor, how many times further does he travel than someone traveling from the first to the fourth floor?"

Note: The teacher should be prepared to give credit to the student who mentions that there are likely many more than 16 "stories" in such an apartment tower; after all, there are eight million stories in New York. A discussion of the possible non-labeling of the thirteenth floor as the thirteenth floor may also be expected.

Typical student thinking for this problem may go thusly: "The first person traveled 16 floors, the second person traveled only four. That's a ratio of 16-to-4, so the first person traveled four times as far." Not so. Here's why:

The first person actually traveled only 15 floors—there was no need to "traverse," so to speak, the 16th floor. Likewise, the second person traveled only three floors. The ratio, therefore, is 15-to-3, so the first person traveled *five* times as far as the second person.

As is often the case, a chalkboard sketch of the situation might be helpful (provided by a student if possible).

The next problem, "The Runner," challenges students to use what they have hopefully learned from the Lift problem.

#17: THE RUNNER

Curriculum Areas: Time, Rate, and Logic

"A runner begins running a road race at a constant pace at mile marker #1 at noon. She passes mile marker #5 at 12:20 P.M. When will she pass mile marker #10?"

Most students will compute that the runner's pace is four minutes per mile, based upon traveling five miles in twenty minutes. Ignoring the impossibility of anyone, male or female, running that fast for that long, this line of thinking misses the fact the mile marker #5 is located only *four* miles into the race. The runner's actual pace, therefore, is a more plausible five minutes per mile. That would put her at mile marker #10 (*nine* miles into the race) at 12:45 P.M.

As with other problems of this type, a sketch would help. The teacher can always offer to put a sketch on the chalkboard once a student has provided an answer; another option is for the teacher to encourage students to make their own sketches at their tables or desks—this will encourage the use of sketch-making as a routine problem-solving technique.

#18: THE HUNGRY BOOKWORM

Curriculum Area: Fraction Addition

"A very tiny bookworm eats his way from Volume I, page 1, to the

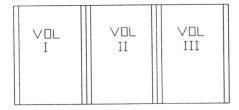

Figure 6.

end of Volume III, the last page in this set of books. Given the assumptions that follow, how far does the bookworm travel as he eats?" (See Figure 6.)

- Assume that the first page inside each volume is page 1.
- Assume that each cover is 1/4-inch thick.
- Assume that the pages of each volume total two inches in thickness.

Students may see this as a straightforward addition problem. Most will provide an answer of seven inches, arrived at by adding together all but the outer two book covers. Seven inches is incorrect, and here's why:

Page 1 of Volume I is not where most youngsters first think that it is. The page which is farthest to the left in the drawing is actually the *last* page of Volume I! That this is true can be easily demonstrated with any real hard cover book. Similarly, the last page of Volume III is not at the far right; it's located two inches to the left.

This, then, is not so much an addition-of-fractions problem as it is one requiring careful thought before any calculations begin. As students encounter problems of a similar ilk in the future, the teacher may want to refer them back to their experience with the little bookworm.

The worm, by the way, traveled three inches.

#19: THE BULB ON THE BICYCLE TIRE

Curriculum Areas: Geometry, Circles, Curves, and Functions

Even most adults have difficulty in choosing the right answer to the Bulb on the Bicycle Tire problem.

"Imagine that a tiny but bright light bulb is affixed to the outer rim of a bicycle tire. (Assume there is no need for batteries or wires.) Imagine, further, that it's nighttime, and that you are watching the bicycle move perpendicularly across your path from some distance away. Which of the paths shown in Figure 7 will the bulb trace out?"

The surprise correct answer is (b). A great way to allow students to discover this is for the teacher to provide pairs or groups of students with circles (paper plates will work well). Then, the teacher should have them each draw a dot at the edge of their circles, and then experiment to determine the correct answer. The teacher should solicit student suggestions regarding possible procedure. *One method:* One student carefully rolls the marked circle along a table top as others, standing several feet away, stare at the dot, trying to note the path it describes. The teacher might suggest that students squint—surprisingly, this will make the path more obvious.

Note: The dots on the paper plates (and the bulb on the bicycle tire) follow a path known as a *cycloid*. The teacher can use this activity as an introduction to the fact that many smooth curves can be *defined*. (Assignment for an interested student: What sort of curve is the St. Louis Gateway Arch, and how can that curve be duplicated with a 20-inch piece of string?)

#20: THE MAGIC EGG BASKET

Curriculum Area: Time

"Imagine a magic egg basket; magic, because the number of eggs in

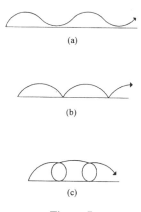

Figure 7.

this very large basket doubles every minute. If one egg was put into the basket at noon, and if the basket was full at 1:00 P.M., when was the basket half-full?"

The first student guess is nearly always 12:30 P.M.—half of the time from noon until 1:00 P.M. If the teacher simply says "no" to those guesses, and provides no further clues, it will only be a few moments before several students see their error. (If the basket was half-full at 12:30 P.M., it would be full at 12:31.)

The correct answer, of course, is 12:59 P.M.

#21: BIRTHDAY MATCH

Curriculum Areas: Probability and Calendar

The answer to the following question may be the epitome of counterintuitiveness. Students will not at first accept the correct answer—only empirical evidence will convince them (suggestions along those lines follow the question).

"What is the minimum number of people who need to be assembled in a room so that the odds are '50/50' that two of them have exactly the same birthday (month and day; not year)? Assume that the people's birth dates are randomly distributed."

The teacher may want to set up the discussion by backtracking for a moment and asking the following question: "How many people would need to be in a room for it to be a certainty that two of them have the exact same birthday?" Many students will quickly be able to provide the correct answer of 366—that's 365, plus one, of course. (367 is the correct answer, if we wish to account for leap year.) With that, students may make a first guess of 183—half of 366.

The teacher can expect other guesses wildly in excess of the correct answer, which is 23 people.

Even students who have had experience with counterintuitive problems, and who have developed a good sense of caution when approaching questions such as these, will state that 23 simply cannot be right. The way to verify the answer, of course, is to have students check several other classrooms in the school for common birthdays. (The teacher should, of course, allow students to come up with this verification suggestion. He may even wish to appear to agree to the proposal reluctantly—as if he knows he might be wrong—for the sake of heightening student motivation.) If a classroom has exactly 23 students that's good, but classrooms with more than 23 will work if some "extra" students are randomly discounted from the start. The odds are very high that common birthdays will be found before the students have checked even three rooms.

Note: The teacher should make sure that students who have volunteered to check the answer of 23 do not make a common mistake. It would be incorrect for the checkers to ask an *individual* student for his/her birthday, and then to ask if there is a match in the classroom. The odds are indeed low that there would be a match. The checkers need to have students call out their January birthday dates, then their February dates, and so on, as a method for finding matches.

The mathematics involved in verifying 23 as the answer are probably beyond the reach of most middle-grade students. Nonetheless, the math looks like this: $1 - [(365 \times 364 \times 363 \times \ldots 343)/365^{23}]$ = the probability that at least two randomly chosen people will share the same birthday (343 is the 23rd whole number counting down from 365).

#22: A MONTH OF SUNDAYS

Curriculum Area: Calendar

"What is the greatest possible number of Sundays in a month?"

Most students will first give an answer of four. The teacher should press students as to why four was chosen. Their reasoning may be along these lines: it takes seven days to make a week, and one can only get four groups of seven days out of 31. Typically, at just about the same time that students are stating that rationale, they see the error in their thinking without any further prompting from the teacher.

A month can contain four groups of seven days *plus a fraction of a group of seven days*. That fraction can contain a fifth Sunday.

Some good follow-up questions for students:

(*1*) Can a month ever contain only three Sundays? If so, what would be the circumstances? If not, why not? (The answer is no. The shortest month has 28 days, and there is no way to "arrange" those 28 days without including four Sundays.)

(*2*) Would changing "Sunday" to "Monday," or to any other day of the week, change the answers of any of these questions?

Extension Activity

The teacher may want to challenge a student or students to research and report on why Easter falls on different Sundays each year. (The answer is that Easter always falls on the first Sunday following the first full moon after the first day of spring.) The report should include an explanation of why the earliest and the latest possible Easter dates can be about five weeks apart. Respecting the need to avoid the teaching of religion in public schools, this is a good opportunity nonetheless to allow a willing student to report on the unusual history of the setting of dates for Easter. The Jewish calendar also sets dates for important holidays, which do not fall on the same Gregorian calendar dates each year, and this provides another fine opportunity for student research. (Actually, the determination of dates for Easter is related to the Jewish lunar calendar!)

#23: HOW LONG?

*Curriculum Areas: **Geometry and Length***

This activity demonstrates that students—most people, in fact—are poor guessers when it comes to physically measuring things that are compacted or whose appearance has been drastically altered in some way.

The teacher should take a ten-foot length of yarn and tape it to the chalkboard or wall in a tight zigzag pattern, as in Figure 8. This should be done before class, or at another time when no students are looking.

Once class begins, the teacher should solicit student guesses regarding the length of the yarn. Invariably, students will provide answers that underestimate the length of the yarn, sometimes drastically so. After students have had a chance to provide various guesses and estimates, the teacher should take the yarn down, and with student assistance, hold it out to its full length. Before allowing a student volunteer to measure the length of the yarn, students should be given the chance to change their guesses. (They will likely increase their estimates of length.)

It is the teacher's job at this point to see to it that the ensuing student discussion centers on the question, "Why do we tend to underesti-

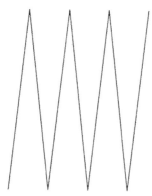

Figure 8.

mate?" If students have ever had the occasion to purchase, say, a coiled two-foot length of telephone cording in a little transparent plastic bag, they can confirm that this tendency to underestimate is real. Those two-foot cords look at first glance as if they might be only one foot long.

A variation of this activity is to provide students with little wadded-up balls of yarn, again asking them to provide estimates of length. Tightly rolling the yarn into a ball (instead of just randomly wadding it up) may further exaggerate student underestimation.

Although student underestimation may still be the norm in these cases, the teacher should expect that students will make better guesses all the time; their guesses are now *educated* ones.

Extension Activities

Teachers may want to consider offering students quick measurement challenges often, as frequently as every day. Helping students become comfortable and familiar with numerous examples of real-life lengths, weights, capacities, and temperatures will benefit them when they undertake formal study of those topics.

Here is one idea for promotion of "measurement literacy": The teacher should offer numerous opportunities for students to make measurement guesses and estimates, with no grades or other evaluation risks attached. When students enter the classroom daily, for instance, they can be asked to jot down on a slip of paper how heavy they think a given weight is—a nice rock borrowed from the science room, perhaps. The teacher can collect these slips, share student answers aloud, and ask class members to offer their opinions regarding the estimates. Another option is to provide students with multiple choice answers on the chalkboard, asking them to choose one (one kilogram? three? ten? etc.). Frequent—even daily—practice will find students becoming better and better guessers/estimators.

Here are a few other daily measurement activity ideas. Students might estimate:

(*1*) The number of pencils (or beans, or grains of rice) in a clear jar
(*2*) The length of various chalk lines drawn on the chalkboard (with students forbidden from getting nearer than, say, six feet from the lines)
(*3*) The temperature of a given large container of water (with a discussion of likely falling temperature to be expected). Students should be allowed to touch the water.
(*4*) The number of pennies stacked on the very edge of a meter stick (extended over a table edge) that will cause the stick to topple
(*5*) The weight of any number of classroom items, including the weight of the teacher or students (willing volunteers only!)

Teachers can add to this short list, or they can ask willing students to brainstorm and produce a list of, say, one hundred measurement estimation activities for the school year.

The phenomena seen in the next activity, "Kount the Korn Kernels," is related to the one at work in "How Long?"

#24: KOUNT THE KORN KERNELS

Curriculum Areas: Volume and Estimation

The teacher should count out several hundred unpopped popcorn kernels, and put them in a clear small bowl or glass. Students should then be afforded the opportunity to guess/estimate the number of kernels and provide rationales for their answers, verbally or in writing. Students will nearly always underestimate the number.

Student tendency to underestimate in this activity can be drastic. It would not be unusual for some students to see 540 kernels as 150 or less.

The teacher can keep this activity very informal, or, she can turn it into a contest. Two suggested contest formats:

(*1*) The teacher can give each student a few moments to look at the kernels (without touching them), then she can ask each student to put his name and best guess on a slip of paper. The best estimate, of course, wins. The winner should then be asked to provide "the secret" of his/her good guess.

(*2*) Individual or small groups of students can be challenged to come up with suggested methods for obtaining good estimates (in writing) without being allowed to touch the kernels. Once written, students can share their suggestions with the rest of the class, and the teacher can decide, or other class members can vote on, the best methodology (followed up with producing an estimate by that method, and then, actually counting the kernels). If no student suggests that visualizing layers of kernels is an aid to good estimation, the teacher should provide that tip. ("It looks like there are about 50 kernels in the bottom layer, and I think that there are about 10 layers.")

The teacher should give students additional practice (the same day and in the days following) in this sort of estimation. Jelly beans and ping pong balls should be considered for follow-up estimation activities. A benefit of this activity is that it sets up students for a more formal study of volume later on. The layering concept is important to eventual student understanding of the $L \times W \times H$ volume formula and its derivatives.

It may be that students' experience with the preceding problem ("How Long?") will have prepared them for better estimation during this activity. If the teacher senses that, indeed, "the lesson wasn't learned," she may again want to emphasize the common measurement underestimation phenomenon.

Extension Activity

Once the teacher has exhausted most discussion possibilities regarding the "korn kernels," she can actually pop the kernels in the school kitchen—but not before soliciting new student predictions regarding the likely volume of the popped corn. Students can also be asked to compute the ratio of unpopped corn-to-popped. (Allowing students to eat the popped corn is a nice way to close out this activity.)

The activity described above can be further extended by a teacher-led discussion of the predicted ratio of snow-to-melted snow.

#25: SOCKS IN THE DARK

Curriculum Areas: Probability and Logic

"Young Freddy's sock drawer contains 10 loose blue and 10 loose black socks, thoroughly mixed together. Early one morning a snowstorm knocks out all power in Freddy's city, but as school is never canceled in his town, he must still get dressed in the morning. He wants to grab enough socks from his drawer (in the dark) to be sure of having at least one matching pair with him as he brings them into the living room, where there is a small, battery-operated lamp. How many socks should Freddy grab to be absolutely sure of having at least one matching pair?"

The first student response will likely be "12 socks."

The answer is just *three* socks.

If any counterintuitive problem is likely to make students literally call out "That can't be right!", it's this one. That insistence generally changes within a few seconds as youngsters see the folly of any answer other than three socks. The teacher can expect several student slaps to the forehead! The first student who sees his mistake should be given the chance to explain the correct answer to the others.

The mental miscue at work here is this: If I want a pair of blue socks, I indeed need to grab twelve individual socks in the dark, because it's just possible that ten of them are black. What this thinking misses is that those ten socks certainly provide Freddy with a matching pair (albeit black), and the problem did not ask for a pair of a specific color.

#26: THEY CAN'T BE EQUAL!

Curriculum Area: Decimal Notation

"Given that part of this short equation is covered up, is it somehow possible that the equation is absolutely true?"

$$1.0 = 0.?\boxed{}$$

The answer (surprising even to many adults) is yes. The complete equation is

$$1.0 = 0.\overline{9}$$

Even after the solution is shown to students (hopefully, a student will have provided it!), many may complain that $0.\overline{9}$ is not *exactly* equal to 1.0; that it is somehow a little "short." There are two ways that the teacher can counter these arguments:

(*1*) The teacher should reinforce the idea that, in decimal notation, the repeating bar over the nine indicates that the series of nines goes on forever, "to infinity," and that as long as the series is infinite, the number is a whole one.

(*2*) The teacher can ask students for the decimal equivalent of one-third. Most will know that one-third is equivalent to $0.\overline{3}$, exactly. Students should then be asked to perform the following addition:

$$0.\overline{3} + 0.\overline{3} + 0.\overline{3}.$$

At this point, many students will "see the light" regarding the fact that 1.0 does indeed equal $0.\overline{9}$!

Teachers' Review of Decimal Notation

The repeating bar in decimal notation (shown here over the 4; $0.\overline{4}$) is used to indicate that the number or block of numbers below the bar continue to repeat indefinitely.

Any value that can be shown as a common fraction (an integer over a counting number, such as 3/5, –2/7, 23/9 or 0/20) can be shown as either a terminating or a repeating decimal. Dividing the numerator by the denominator will provide the decimal.

These fractions convert to the terminating decimals shown:

$$\frac{2}{5} = 0.4 \qquad \frac{7}{100} = 0.07 \qquad \frac{3}{8} = 0.375$$

The following can be shown as repeating decimals:

$$\frac{2}{3} = 0.\overline{6} \qquad \frac{1}{7} = 0.\overline{142857} \qquad \frac{5}{11} = 0.\overline{45}$$

Can it be predicted whether a fraction will be a terminating or a repeating decimal? Yes. If the prime factorization of the denominator contains only twos and fives, the decimal equivalent will terminate. All others will repeat. Examples:

1/2, 3/5, 7/8, 3/10, 15/16, and 147/200 will convert to terminating decimals.

1/6, 5/7, 7/9, 5/11 and 13/51 will convert to repeating decimals.

#27: THE LAST DAY

Curriculum Area: Calendar

"What will be the last moment of the 20th century?"

"The end of the day, December 31, 1999" is the response that the teacher can typically expect from students. In recent years the media have helped to establish in the public consciousness that the century and the current millennium will end at the close of the last day of 1999. Certainly there is a fascinating novelty inherent in seeing the thousands-place change (to a 2) for the first time in a thousand years, and it is difficult for any mathematician or teacher to counter the notion that the millennium is over on that date.

In fact, the last day of the century (and the millennium, and the decade) will be December 31, 2000. Why? Because the year known as 2000 is the one-thousandth year of the millennium, and the one-thousandth year needs to be *over* before one can say that a new millennium has begun. Think of it this way: at the end of 1999, exactly 1,999 years of the millennium have gone by, *not* 2,000 years. January 1, 2000 marks the beginning of the one-thousandth (and last) year of the millennium.

#28: I'M MY OWN GRANDMOTHER!

Curriculum Area: Logic

Can someone be her own grandmother?

For the sake of discussion, let's agree on the following definitions:

(1) A person's mother-in-law and father-in-law can be called the person's mother and father (that is, we can ignore the designation of "in-law").

(2) Anyone married to someone's father is her mother; anyone married to someone's mother is her father (that is, ignore the designation of "step").

I'm My Own Grandmother! 49

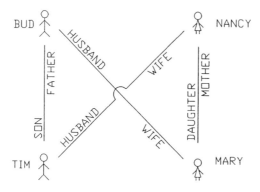

Figure 9.

(3) Anyone's father's father is her grandfather. Anyone's father's mother is her grandmother. Likewise, anyone's mother's mother is her grandmother, and anyone's mother's father is her grandfather.

(4) Anyone married to a grandfather is a grandmother (and vice-versa).

We're assuming, too, for the sake of discussion, that there may have been previous marriages for some of these people.

Teachers need to be alerted that this is not a problem to be presented in the closing moments of a class hour, with a quick solution expected. Students need to be given considerable time to work out the possibilities here. Many will benefit from a teacher suggestion that they use sketches of male and female stick figures as problem-solving aids. ("Making a sketch" is a primary problem-solving strategy.)

If this problem is presented to the entire class, and if the teacher encourages a controlled free-for-all discussion, that's fine, but here is one caveat: more aggressive students are likely to dominate the discussion. That's why the teacher may want to consider assigning this problem to individuals or to small groups of students to be wrestled with outside of class.

Figure 9 shows the relationship among four adults that allows us to say that the younger female is her own grandmother.

Further explanation:

- Nancy is Mary's *mother*.
- Whomever is married to Nancy is Mary's *father* (according to our agreed-upon definitions).
- Nancy is married to Tim, therefore he is Mary's *father*.
- Bud is Tim's *father*, and, by marriage, Bud is *Nancy's father*, too.

- By definition, Nancy's father is Mary's *grandfather*.
- Bud, therefore, is Mary's *grandfather*.
- Mary is married to Bud (her grandfather), therefore, *Mary is her own grandmother*!

Obviously the legitimacy of this whole affair is dependent upon acceptance of the definitions. In any case, it provides for a great challenge and a lot of fun.

#29: THE THREE DOORS[1]

Curriculum Area: Probability

In 1990 and 1991, *Parade* (a supplement to many Sunday newspapers) published a series of columns written by Marilyn vos Savant, who is listed in the "Guinness Book of World Records Hall of Fame" for "Highest IQ." In these columns, Ms. vos Savant presented a mathematics/logic problem that caused a good bit of controversy nationwide. Her "Three Doors" problem looked simple on the surface, and its answer seemed obvious. The answer that she put forward, however, met with enormous opposition from her readers, who included PhDs from across the nation. She quoted some of those math experts in one of the columns:

> You are in error—and have ignored good counsel—but Albert Einstein earned a dearer place in the hearts of the people after he admitted his errors. (*from a PhD at Michigan State University*)

> I have been a faithful reader of your column and have not, until now, had any reason to doubt you. However, in this matter, in which I do have expertise, your answer is clearly at odds with the truth. (*from a Millikin University PhD*)

> May I suggest that you obtain and refer to a standard textbook on probability before you try to answer a question of the type again. (*from a PhD at Georgia State University*)

> You are utterly incorrect about the game show question, and I hope this controversy will call some attention to the serious national crisis in mathematics education. If you admit your error, you will have contributed to-

---

[1] The Three Doors problem is reprinted with permission from *Parade* and Marilyn vos Savant, copyright © 1990 and 1991.

ward the solution of a deplorable situation. How many irate mathematicians are needed to get you to change your mind? (*from a PhD at Georgetown University*)

Maybe women look at math problems differently than men. (*from a reader in Sunriver, Oregon*)

You're wrong, but look at the positive side. If all those PhDs were wrong, the country would be in serious trouble. (*from a U.S. Army Research Institute PhD*)

As Ms. vos Savant stated in one of those columns, "When reality clashes so violently with intuition, people are shaken."
Here is the problem that caused such a stir:

"Suppose you're on a game show, and you're given a choice of three doors. Behind one door is a car; behind the others, goats. You pick a door—say, No. 1—and the host, who knows what's behind the doors, opens another door" say, No. 3—which has a goat. He then says to you, "Do you want to pick door No. 2?" Is it to your advantage to switch your choice?

The intuitive answer is "No, switching is not to my advantage. The odds are equal that the car is behind door 2 or door 3." Intuitive, but incorrect. It is indeed to the player's advantage to switch. The odds that the car is behind door No. 2 are *twice as good* as the odds that it's behind door No. 1. Here is one explanation:

If the host had originally said to the player (you), "You may pick one door, and what's behind it is yours to keep, or, you may pick a set of *two* doors, and keep what's behind both of them," what would you do? You'd pick two doors, naturally, doubling your chances of getting the car. Well, that's really what the host is offering you when he asks you if you'd like to switch! You've picked door No.1, which probably doesn't hide the car, and the host is in effect offering you the set of doors 2 and 3, one of which probably does hide the car, and he's done you the *favor* of telling you that it's not behind door No. 3. That is, the car is probably behind door 2 or 3, *and it's definitely not behind No. 3!* Of course you should switch to door No. 2!

In the face of continuing opposition to this strongly counterintuitive answer, Ms. vos Savant offered to put her answer to the empirical test. She asked mathematics teachers across the country to set up a little probability experiment with their students replicating the conditions of the Three Doors problem, and to send in the results to her at *Parade*. (The experiment is outlined below.) Those results confirmed that she was correct. If middle-grade mathematics students wrestling with this problem remain unconvinced of the correctness of the answer, Ms. vos Savant's experimental guidelines are reprinted here:

> One student plays the contestant, another plays the host. Label three paper cups No. 1, No. 2, and No. 3. While the contestant looks away, the host randomly hides a penny under a cup by throwing a die until a 1, 2, or 3 comes up. Next, the contestant randomly points to a cup by throwing a die the same way. The host lifts up a losing cup from the two unchosen. Last, the contestant "stays" and lifts up his original cup to see if it covers the penny. Play "not switching" 200 times and keep track of how often the contestant wins.
>
> Then test the other strategy. Play the game the same way until the last instruction, at which point the contestant instead "switches" and lifts up the cup *not* chosen by anyone to see if it covers the penny. Play switching 200 times also.

Conducting this experiment (there is no need to try each version 200 times; 50 times will be sufficient) will verify the answer.

Another interesting method for verifying Ms. vos Savant's answer is by modeling the experiment on computer. Although this may be too advanced for most middle-grade youngsters, the following computer program is offered nonetheless.

```
10 REM*THIS PROGRAM IS WRITTEN BY LISA ROMBES
20 REM*MAY 8, 1997
30 REM*TO SIMULATE THE "THREE DOORS" PROBLEM
40 CLS
43 PRINT "YOU ARE ON A GAME SHOW WITH THREE DOORS"
45 PRINT "YOU ALWAYS SELECT DOOR #1. THE CAR IS EITHER IN"
47 PRINT "DOOR # 1, 2, OR 3."
50 PRINT "THIS PROGRAM SIMULATES THE RESULTS OF"
60 PRINT "SWITCHING DOORS AFTER BEING SHOWN DOOR WITHOUT THE CAR"
70 PRINT "HOW MANY TRIALS DO YOU WANT TO RUN?"
80 INPUT "TYPE A NUMBER AND PRESS RETURN"; X
90 FOR NUM = 1 TO X
100 LET Y = (INT(RND(1) * 3)) + 1
102 PRINT
```

```
105 PRINT "CAR IS BEHIND DOOR NUMBER"; Y
110 IF Y = 1 THEN PRINT "PICK # 1, SWITCH AND LOSE":
GOTO 135
120 IF Y = 2 THEN PRINT "PICK # 1, SHOWN # 3,": LET
COUNT = COUNT + 1
130 IF Y = 3 THEN PRINT "PICK # 1, SHOWN #2,": LET
COUNT = COUNT + 1
132 PRINT "SWITCH AND WIN": PRINT
135 FOR DUMMY = 1 TO 310: NEXT DUMMY: REM*TIME-
DELAY LOOP
140 NEXT NUM
150 PRINT "YOU PLAYED THE GAME"; X; " TIMES."
155 PRINT "YOU SWITCHED EACH TIME"
160 PRINT "YOU WON"; COUNT; "TIMES,"
170 PRINT "YOU LOST"; X - COUNT; " TIMES."
180 END
```

#30: ODDS ON TWO BOYS[2]

Curriculum Area: Probability

As if readers of Ms. vos Savant's magazine column were not convinced of her expertise (see problem #29: The Three Doors), another equally hot controversy erupted in 1997 when she presented another problem, asking again for reader responses. It, too, is an appropriate one for use with middle-grade youngsters. And, like the preceding "Three Doors" problem, it is highly counterintuitive. Here is the original problem:

> A man and a woman (unrelated) each have two children. At least one of the woman's children is a boy, and the man's oldest child is a boy. Do the chances that the woman has two boys equal the chances that the man has two boys?

Ms. vos Savant offered that the chances that the woman has two boys is 1 in 3, and that the chances that the man has two boys is 1 in 2,

[2]Odds on Two Boys problem is reprinted with permission from *Parade* and Marilyn vos Savant, copyright © 1997.

that is, that the chances were indeed not equal. Again, most readers who responded to her answer disagreed with her. Some of their comments to her included these:

> I will never read your column again. (*from a reader in Tucson, Arizona*)
>
> I can only conclude that you are not woman enough to face the truth and admit your mistake. You are highly intelligent, and that is an admirable quality, but high intelligence coupled with an unwillingness to admit a mistake is unforgivable. (*from a reader in Overland Park, Kansas*)
>
> This is not going to go away until you admit that you are wrong, wrong, wrong! (*from a reader in Salt Lake City*)
>
> You are not the only genius to base logic on a faulty premise. Einstein did it more than once. (*from a reader in North Myrtle Beach, South Carolina*)

One of Ms. vos Savant's early supporters was a person with a PhD in nuclear engineering; the following excerpts from a detractor's letter refer to that engineer:

> I was horrified to read that one of your few supporters was an engineer responsible for assessing risks in the operation of nuclear power plants. I sometimes wonder why critics of IQ testing don't point to some of your work as vivid examples of the vast difference between IQ and logic. (*from a reader in Knoxville, Tennessee*)
>
> I guess the real hope is that the nuclear engineer wasn't paying very close attention when she offered her assent, or else the next problem will involve three-eyed children. (*from a reader in White Bear Lake, Minnesota*)

Middle-grade math teachers are likely to get similar responses from students. The mistake likely to be made by students (and by most of Ms. vos Savant's readers) is this: Having two children where at least one is a boy, and having two children where the oldest is a boy *are not identical situations*. An explanation follows.

For anyone "matching" the woman's situation (two children; at least one is a boy), there exist exactly three equally likely possibilities: older child boy/younger child boy, older child boy/younger child girl, or, older child girl/younger child boy. For anyone "matching" the man's situation (two children; older child a boy), there are exactly two equally likely possibilities: older child boy/younger child boy, or, older child boy/younger child girl. The problem's "assigning" of the man's "original boy" to the older age slot suddenly limits the man's family possibilities a bit, a limitation not present for the woman, whose "original boy" can be older or younger.

Note: The "Odds on Two Boys" problem is a variation of a problem first presented, apparently, in *Mathematical Know-How*, by the Russian Boris Kordensky. His book (the English version is the famed *The Moscow Puzzles*) is a highly recommended collection of hundreds of classic and original problems, many suitable for middle-grade students.

CHAPTER 3

Challenges and Puzzles

#31: TRUTHTELLER OR LIAR?

Curriculum Area: Logic

THIS OUTSTANDING CLASSIC logic problem cannot be strictly classified as mathematical; but the teacher will see that the very same "brain cells" that allow a student to solve an algebraic equation are at work here.

"A stranger is visiting the faraway island of Trutherlie. He is walking down a path when he comes to a fork in the road (see Figure 10). He sees two natives on the two paths, and he knows that one is of the Truthteller tribe, and the other is a member of the Liars tribe (the only two groups living on the island). The visiting stranger does *not* know which man is from which tribe, but wishes to know. (Need it be mentioned that Truthtellers *always* tell the truth, and Liars *always* lie?)

The visitor asks the man on his left this question: 'Are you a Truthteller?' The man mumbles an answer which the visitor cannot quite hear. The man on the visitor's right (wishing to be helpful) then loudly announces, 'He said yes,' (referring to the man on the left).

The visitor now has enough information to determine which man is the Truthteller, and which is the Liar. Which is which, and how do you know for sure?"

Students will naturally employ various versions of the "what if?" method of problem solving—very appropriate here! Only after most of them have exhausted their mental resources in trying to solve this one should the teacher offer clues in the form of the following questions:

(*1*) "If the man on the left was a Truthteller, what did he say in answer to the visitor's question?" (He would have said "Yes.")

57

58 CHALLENGES AND PUZZLES

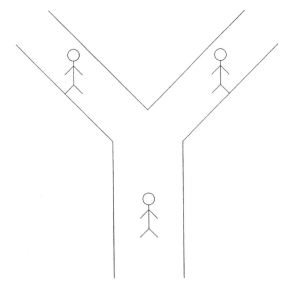

Figure 10.

(*2*) "If the man on the left was a Liar, what did he say in answer to the visitor's question?" (He also would have said "Yes.")

(*3*) "Then what did the man on the left say, *regardless of his tribe of origin?*" (He said "Yes.")

(*4*) "If the man on the right *claimed* that the other fellow said "Yes," (and we know this to be true), then what does that make the man on the right?" (A Truthteller, of course!)

So, the man on the right is a Truthteller; the man on the left, a Liar.

#32: TWO TRAINS, ONE UNFORTUNATE FLY

Curriculum Area: Rate

"Train A leaves Poughkeepsie for New York City at the same time that Train B leaves New York for Poughkeepsie (3:00 P.M.). They travel toward one another on the same track, traveling at a consistent speed of 60 mph. (Poughkeepsie and New York are 60 miles apart.)

Imagine that there is a fly on the very front of Train B as it leaves New York. Imagine further that the moment the train begins moving, the fly flies toward Train A at a consistent speed of 140 mph. When the fly reaches that other train, it immediately turns around and begins flying back toward the Poughkeepsie-bound train. Imagine that each time the fly reaches a train, it turns around and flies toward the other one.

The question is this: When the two trains meet, *how far will the fly have traveled?"*

The teacher should expect the following questions from students before they attempt to tackle the mathematics involved. "Wouldn't the trains crash?" and "Does the fly get squashed?" To facilitate, the teacher may want to answer those questions with these answers, respectively: "Yes," and "Yes."

Students will try to solve this problem by sketching an elaborate back-and-forth path for the fly, noting that each train has moved forward since the fly was last there. Although it is possible to answer the question using that complicated method, at some point the teacher will want to offer the following questions as clues for solving this puzzle in a simpler and more elegant fashion:

(1) "When the trains meet (crash), for how long will they have been traveling?" (one half-hour)

(2) "When the trains meet, for how long will the fly have been traveling?" (one half-hour)

(3) "If the fly was traveling at a constant speed of 140 mph for one half-hour . . ."

At this point, someone in the classroom will "see the light." If the fly was traveling at 140 mph for one half-hour, it traveled 70 miles.

This problem allows the teacher to emphasize the need to look for alternative methods of solution when the first method that comes to mind is too difficult.

#33: LET'S PLAY WEATHERPERSON

Curriculum Areas: Time, Percent, and Logic

"It is raining at midnight. Which of these probably represents the chances that we will have sunny weather in 96 hours?

                0%   25%   50%   75%   100%"

Middle-grade students do not generally step back from problems to gain a broad perspective; they are often too close to a problem to see obvious solutions. The teacher will enjoy waiting for the first student who notices that 96 hours is exactly four days, and that sunny weather is unlikely at midnight! (*Note:* If a student mentions that the sun can indeed be shining at midnight during certain times of the year above the Arctic Circle, the teacher should certainly acknowledge the truth of that statement. That is exactly the type of student observation that ought to be encouraged.)

#34: THAT'S JUST PERFECT

Curriculum Area: Factors

The ancient Greeks were fascinated with numbers; indeed, they saw individual numbers as having personalities all their own. For instance, some numbers were considered "perfect." They considered perfect numbers as those whose proper factors add up to the given number. Six is perfect because its proper factors, 1, 2, and 3, add up to 6.

"There is only one perfect number between 7 and 50. Can you find it?"

As students test a variety of numbers for "perfection," they may notice that some numbers' proper factors add to a sum which is *less* than the given number. These were named *deficient* numbers by the ancient Greeks. *Abundant* was the name given to numbers whose proper factors add up to a sum *greater* than the given number. The teacher may want to ask students to identify abundance and deficiency as they work on trying to locate that perfect number.

A simple extension activity is for the teacher to ask students to quickly guess whether a number is abundant or deficient, based strictly on the "feel" of the number. Most youngsters, for instance, will be able to state without calculation that 24 is *abundant*. They probably have a sense that "lots of things go into 24," and this is correct. That 37 is *deficient* may also be somewhat obvious to them. ("Almost nothing goes into 37.")

The perfect number that the students are looking for is 28 (1 + 2 + 4 + 7 + 14 = 28). There have been, by the way, very few other discoveries of perfect numbers.

Teachers' Review of Factors

Factors of whole numbers are generally any other whole numbers that evenly divide into the given number. For example:

- Nine is a factor of 72 because 72 ÷ 9 = 8, with no remainder.
- One and the number itself are always factors of a given number. Some whole numbers have only those two factors. These are the prime numbers.
- Thirteen is prime because its only factors are 1 and 13.
- Non-prime whole numbers (excluding zero) are called composite numbers. All composite numbers have at least three different factors. Zero is neither prime nor composite.
- Thirty is composite because it has more than two factors (1, 2, 3, 5, 6, 10, 15, and 30).

The following systematic method can be used to identify most factors of many numbers. Of course, if the number is very large, the method becomes unwieldy.

To find all factors of, say, 40:

(1) List 1 and the number itself (1, 40).

(2) If the number is even, list 2 (1, 40, 2).

(3) If it's even, take half of it (1, 40, 2, 20).

(4) Try to take half of any new factor, add it to the list (1, 40, 2, 20, 10, 5).

(5) Check to see if 3 and 5 are factors (3 is not in this case; 5 is).

(6) Make sure that all of the listed factors have "partners." (This would add 4 and 8 to the list.)

The factors of 40 are 1, 2, 4, 5, 8, 10, 20, and 40.

This method will "miss" the factors of those more rare numbers that are built only of primes, 1, and the numbers themselves. Ninety-one, for instance, has factors of 1, 7, 13, and 91. The method, however, is useful for finding most factors of commonly used numbers.

An additional note regarding prime numbers: The Greek mathematician Euclid (specialty: geometry) proved that there is no highest prime number. To get a "new highest" prime number, one merely needs to multiply all of the prime numbers, 2 through x (where x = some high prime number), then add 1 to the product. (Or . . . the result will contain a factor which is a "new highest" prime number.)

#35: COMPLETE THE SEQUENCE

Curriculum Areas: Sequencing and Functions

Middle-grade students enjoy the challenge of trying to complete or continue patterned sequences. These activities provide great prompts for appropriate free-for-all classroom discussions. The teacher may want to leave it up to the rest of the class members to decide if another student's suggested solution is acceptable or not.

Here are some strictly numerical sequences:

4 6 8 9 10 12 14 15 __ __ __ (These are the composite numbers, or non-primes; the next three numbers are 16, 18, and 20.)

1 4 9 16 25 36 49 __ __ __ (These are the squares of the counting numbers; the next three numbers are 64, 81, and 100.)

Here are two non-numerical sequences:

O T T F F S S E __ (An "N" should go in the blank; these are the first letters of the words one, two, three, four, etc.)

This sequence should be accompanied by the following question in writing, perhaps on the chalkboard: "What is the next letter in this sequence: W I T N L I T __?" (The answer is an "S"; these are the first letters of the words in the instructions.)

The teacher may wish to challenge students to come up with their own incomplete sequences for other students to try their hands at.

#36: CONVERT IF YOU CAN

Curricular Area: Formulating Rules

"What is the rule for converting the first number to the second? The rule is the same for all examples.

14	820224
19	831459
22	448202
40	088640
9	819099
11	224606"

This is another excellent brainstorming prompt. Allowing students to choose to work individually or in small groups is appropriate here.

The rule is this: Take the original number and double it, then transpose the digits of that product. This gives the first two digits of the second number. Next, add the first two digits, take the units digit of the sum; this gives the third digit. Similarly, add the second and third digits, again take the units digit of the sum; this gives the fourth digit. Repeat this pattern until the sixth digit is produced.

#37: THE PERFECT SOLITAIRE DECK

Curriculum Area: Logic

Here is an unusual challenge for students who are easily bored.

Most students know how to play the card game of Solitaire (many will know both the card and computer versions). The teacher should challenge a student to arrange for him and the class a perfect solitaire deck. Such a deck, when dealt out in the normal solitaire fashion, would produce an arrangement of cards that immediately begin to allow cards to be stacked onto the aces. Every move would reveal the "perfect" next card; the game would be the most efficient game of solitaire possible.

Obviously, if a student claims to have arranged such a deck, no one should be allowed to shuffle the cards before the demonstration begins.

This is a difficult task, but a high-interest activity. Any student who is able to meet the challenge should be singled out for high praise. Naturally, the teacher will want the student to explain his/her methodology to class members. The teacher should consider allowing the student to "take the show on the road," so to speak, by letting him/her provide a demonstration to other classes, or even to the principal of the school.

A good follow-up to this activity is to allow other students (and perhaps the teacher herself!) to demonstrate other card tricks for the class. Many card tricks work because of the mathematics involved—challenging the rest of the class to figure out how the tricks may have been done could be rewarding.

#38: ROTATE THE TRIANGLE

Curriculum Areas: Geometry and Triangles

"The six pennies shown in Figure 11 form an equilateral triangle. By moving exactly two of the pennies to a new position, can you turn the triangle upside-down?"

Students should be allowed to attempt the solution to this problem with actual pennies. This will help to bring in kinesthetic learners. (Some youngsters may rather sketch little circles instead of using real coins. The teacher should take mental notes on this—the activity pro-

64 CHALLENGES AND PUZZLES

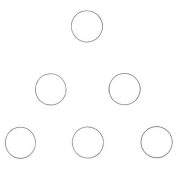

Figure 11.

vides clues as to which students may not prefer or even need manipulatives to solve most problems.) Figure 12 shows the solution.

Middle-grade students rarely solve this problem quickly. Shouts of "This can't be done!" will likely be heard before anyone announces "I've got it!" A worst-case scenario will have the teacher eventually telling the class that the "outer two" pennies are the ones that need to be moved.

It should be announced in advance that no one who believes that he has found the solution should plan to share it with the class immediately. The teacher may want to state that proposed solutions will be shared only after, say, three students have discovered it.

Once several students discover the solution, the teacher should point out that both figures (the original and the inverted triangles) can be described as three-penny triangles surrounded by larger (and similar and inverted) three-penny triangles. Further, the teacher should encourage all students to physically make the transformation themselves several times before concluding the activity.

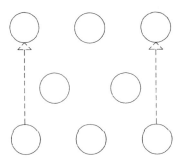

Figure 12.

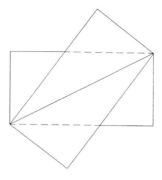

Figure 13.

#39: THE SHRINKING DOLLAR BILL

Curriculum Areas: Geometry, Rectangles, and Diagonals

Although this is what once would have been called a "parlor trick," it helps to reinforce some important geometry vocabulary, and sharpens students' visual/spatial perception.

The teacher should align two dollar bills as shown in Figure 13; note that the equal-length diagonals of each of the bills are perfectly aligned. The bills should then be separated, and the teacher should then challenge individual students to simply duplicate the arrangement that they just saw.

Few students, if any, will be able to do it. Most will align the bills with one bill's diagonal aligned with the other's side.

If the teacher is inclined to have a little more fun with his students, he may want to insist that he has indeed somehow caused one bill to shrink in size. Students may put forward the theory that he has "crinkled" one of the bills in order to shorten its length. Challenging students to come up with an explanation by the next class meeting provides for a high-interest optional assignment.

#40: THE SNOBBY NEIGHBORS

Curriculum Areas: Geometry and Paths

"Figure 14 is a birdseye view of a very exclusive—and snobby—subdivision of four very expensive homes. A, B, C, and D designate the Anderson, Brown, Carter, and Davis residences. The surrounding rectangle represents a fancy brick wall, with the four breaks in the wall representing exits from the subdivision. (Note that the Browns have built a wall from the back of their home to the back wall of the subdivision.)

How snobby is this subdivision? So snobby that each family has its own designated exit as well as its own driveway from home to the exit.

Can you connect each home to its exit by drawing four paths? Remember, this is such a snobby subdivision that the four families won't even cross each other's paths, and neither can you as you attempt to solve this puzzle."

Early on, many students will claim that this puzzle "can't be done." If they are allowed to stay with the task, that will give way soon enough to the realization that they need to curl behind at least some of the houses in order to be successful. The teacher will note that the paths in the solution drawing (see Figure 15) are very close together—the solution is neither elegant or realistic. The teacher should not accept student solutions that are difficult or impossible to decipher.

This is an "unrealistic" problem, in that no one would ever be likely to have to find a solution to a problem just like this in real life. The

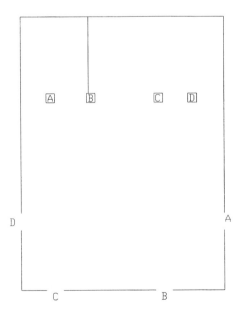

Figure 14.

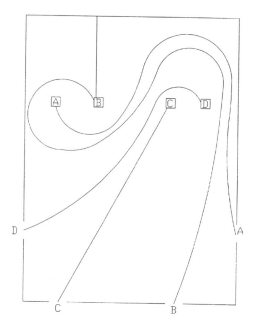

Figure 15.

teacher may want to ask the students, "What value is there in working with "unrealistic problems such as this one?"

#41: IT LOOKS SIMPLE

Curriculum Areas: Square Roots, Factors, and Products

"What is the square root of $\sqrt{1,296}$?"

Some students will reach for their calculators, others' pencils will fly, some may attempt to determine the answer by way of mental calculation. All will likely give an answer of 36, which is incorrect.

The teacher should simply affirm that the answer of 36 is incorrect, and wait for someone to spot the catch. The catch is, of course, that the problem reads, "What is the square root of the square root of 1,296?" The square root of 1,296 is 36, and the square root of 36 is 6.

A good follow-up to this problem is a look at some methods that students can use for determining some square roots mentally.

Let's look at $\sqrt{1,296}$. It ends in a 6, meaning that the square root of 1,296 must end in either a 4 or a 6. (Why? Because a number which ends in 4, squared, will always give a number that ends in 6. Likewise, a number that ends in 6, squared, will always give a number ending in 6.) Middle schoolers can probably see mentally that 30^2 is 900, and that 40^2 is 1,600, meaning that the square root of 1,296 (1,296 being between 900 and 1,600) is between 30 and 40. Given that $\sqrt{1{,}296}$ must end in a 4 or a 6, $\sqrt{1{,}296}$ must be either 34 or 36. (At this point, actual multiplication will confirm the correct answer of 36.)

All of the following rules are based on the assumption that the number in question is a whole number, and that it has an integral square root. Remember, most whole numbers do not have integral square roots (square roots which are themselves whole numbers).

Students should learn the following facts as aids to determining many square roots mentally:

- The square root of any number ending in a 0 will end in a 0.
- The square root of any number ending in a 1 will end in a 1 or a 9.
- The square root of any number ending in a 4 will end in a 2 or an 8.
- The square root of any number ending in a 5 will end in a 5.
- The square root of any number ending in a 6 will end in a 4 or a 6.
- The square root of any number ending in a 9 will end in a 3 or a 7.

(*Note:* Whole numbers ending in a 2, a 3, a 7, or an 8 have no integral square roots.)

In order to quickly estimate square roots, students should also memorize certain "benchmark" squares:

$$10^2 = 100$$
$$12^2 = 144$$
$$15^2 = 225$$
$$18^2 = 324$$
$$20^2 = 400$$
$$25^2 = 625$$
$$30^2 = 900$$
$$40^2 = 1{,}600$$
$$50^2 = 2{,}500$$

Armed with these bits of hard knowledge, students should be able to determine many square roots in their heads. For example: "What is the square root of 1,849?" Given the facts noted above, the square root of 1,849 must end in a 3 or a 7 (40 × 40 = 1,600, so the square root is probably in the 40s). The answer, therefore, is either 43 or 47. Because 1,849 is fairly near 1,600, our best bet is 43.

Here are some numbers that provide good square root–finding practice:

$$361 \quad 484 \quad 841 \quad 1{,}225 \quad 1{,}681$$

#42: CONNECT THE DOTS

Curriculum Areas: Geometry and Paths

This is a classic problem. The lesson, as students will see after completing the activity, is that *one should not assume restrictions* if none are given.

"Can you connect the nine dots in Figure 16 using only four straight segments, drawn continuously?"

The teacher should allow students ample time to experiment with possible solutions, perhaps as a sponge activity, or as a challenge to be done outside of the classroom.

It's desirable that a *student* first bring up the fact that the solution required moving outside of the square formed by the nine dots. As always, it is hoped that a student or students will be able to present a solution to the entire class. The teacher should point this out only if no student does so during guided discussion.

Figure 17 shows the solution.

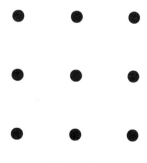

Figure 16.

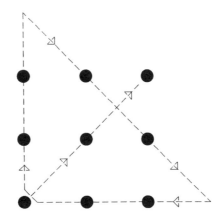

Figure 17.

#43: TWO AMERICAN COINS

Curriculum Area: Money

"I have two standard, modern, legal-tender American coins in my pocket. They total $0.55. One is *not* a nickel. What two coins do I have in my pocket?"

The answer: a half-dollar and a nickel. (One is not a nickel, but the other is.)

To increase the drama of the moment, the teacher may want to actually have the two coins in her pocket. Bringing them out for all to see—wordlessly—will help to bring out those desired protests before students realize the catch here.

#44: INTO THE FOREST

Curriculum Area: Logic

Here is a great old "groaner":
"How far into a forest can a dog walk?"
The answer to this classic riddle is, of course, "halfway." (If the dog travels any farther in the same direction, it is no longer walking into a forest, but out of it.)

Arguments regarding the definition of "into" may ensue; this is a good thing.

#45: THE FROZEN CLOCK

Curriculum Area: Time

"In what way is a clock frozen at 2:24 more accurate than a clock that's permanently three minutes slow?"

Soliciting students' thoughts regarding this question by way of a written response makes for a wonderful assignment. The teacher may ask individual students to provide a short paper that attempts to answer the question, or the student may be given the option of making an oral report to the class. Small group collaboration, with one or two students being assigned the jobs of reporters, also works well.

How is the frozen clock more accurate than the perpetually slow one? It's exactly right twice per day; the other clock, never!

#46: COUNT THE RECTANGLES

Curriculum Areas: Geometry and Rectangles

"How many distinct rectangles can you outline in Figure 18? Remember that they might be overlapping."

Invariably, when first presented with this sort of problem, middle-grade students will underguess. The figure, for instance, contains thirteen distinct rectangles; students are likely to guess between seven and ten at first crack. Even if students move beyond guessing (not that there's anything wrong with that approach as a start!) and attempt to

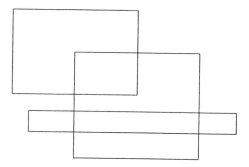

Figure 18.

count the rectangles according to a plan, they will generally miss some. Typically, students will miss some of the smaller rectangles formed by the intersecting sides of larger ones.

These rectangle-counting (or triangle-counting) challenges require no preparation on the part of the teacher; they make, therefore, great sponge activities. The teacher needs merely to sketch a problem on the chalkboard, and the stage is set. It is not necessary for the teacher to know the answer in advance, he can struggle along with his charges. Students will be delighted if the teacher commits to an answer, only to be proven wrong!

Students will get much better at these types of problem very quickly if given enough chances to try them, and their spatial discrimination will be improved.

This activity allows the teacher to review the definitions of rectangle and triangle within a context of problem solving.

Teachers' Review of Polygons

- A polygon is a closed curve made up of three or more straight sides.
- A quadrilateral is a polygon with four sides.
- A parallelogram is a quadrilateral wherein opposite angles have equal measure and opposite sides are of equal length. (When that is true, opposing angles will be of the same length.)
- A rectangle is a parallelogram with four right angles. (When that is true, opposing sides will be of equal length.)
- A square is a rectangle with four equal sides. (All squares are rectangles; not all rectangles are squares.)
- A rhombus is a parallelogram with four equal-length sides. (A "diamond" shape.) All squares are rhombi; not all rhombi are squares.
- A trapezoid is a quadrilateral having exactly two parallel sides.

Other important related terms are these:

- A vertex is a corner of a polygon.
- Adjacent sides are those that are "next to" each other; they share a vertex.
- A diagonal connects vertices of polygons.

#47: THE COUNTERFEIT COINS

Curriculum Areas: Money and Logic

"Here are ten stacks of ten silver dollars each (see Figure 19). All of the dollars are real and identical except for one whole stack, which is composed of ten counterfeit coins. (You can't tell by looking at them, or by handling them, that the fake coins are fake.) Let's assume that all of the real silver dollars weigh exactly 100 grams each. Let's also assume that the counterfeit coins each weigh 101 grams. Using the scale provided (see Figure 20), how can you determine which stack of coins is counterfeit in one weighing? (Stack 1? Stack 2? etc.) By one weighing, we mean that one mix of coins may be laid in the pan all at once, the total weight noted, and that's that. No individual coins or smaller groups of coins may be added or taken away from the pan once the one weighing has been made."

Most early student guesses will contain the misstep of adding coins to the scale pan piecemeal. Naturally, if one could drop individual coins into the pan while reading the scale, one would merely have to wait until a deposited coin increased the total weight by 101 grams to know which pile was counterfeit. The rules of the puzzle disallow this approach.

If students appear stumped after attempting this one for several minutes, the teacher can provide this hint: "Whatever you put into the pan all at once, it must provide a way to discriminate amongst the ten piles. How will one weighing allow the scale to tell you that the stack of fakes is, for instance, stack 4, or 1, or 7?"

Figure 19.

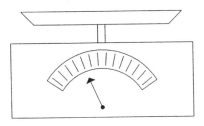

Figure 20.

Should the student frustration level get dangerously high, the solution is hereby provided: One can gather one coin from stack #1, two coins from stack #2, three from #3, and so on through stack ten. Now, all of those gathered coins can be put onto the scale pan at once.

Let us think for a moment. How many coins are we putting into the pan all at once? We're putting in $1 + 2 + 3 \ldots + 10$ coins, or 55 coins. If all of those 55 coins were real and non-counterfeit, their total weight would be 55×100 grams, or 5,500 grams, and that is what the scale would show. But what if exactly *one* of those 55 coins were counterfeit? A fake coin weighs 1 gram more than a real one, so the scale would read 5,501 grams. And that would mean that stack #1 was all counterfeit, because one coin from stack #1 was in the mix. If the scale were to read 5,507 grams, it would be clear that stack #7 was fake—we put in seven counterfeit coins from that stack. Thus, we have a method for determining the counterfeit stack with just one weighing.

Extension Activity

Traveling the path needed to solve this problem, students need to add up all of the whole numbers from one to ten. Although that particular short set of addends is easy to handle, there is a shortcut formula (aren't *all* formulas shortcuts?) for adding together a series of consecutive whole numbers which begins with 1:

Where n is the end number of a sequence, $n(n + 1)/2$ gives the sum

In the counterfeit coin problem: $[10(10 +1)]/2$ gave us 55.

Interested students may want the assignment of determining the sums of various series (1–20? 1–500? 1–2,799?).

#48: A STREETCAR NAMED DECEPTION

Curriculum Area: Mental Arithmetic

The teacher should give this problem verbally to a student volunteer, asking him to listen carefully, and to be prepared to answer a question at the conclusion.

"A streetcar in San Francisco with 13 people on board makes a stop: 3 people get off of the car, while 9 more get on. At the next stop, 5 get off and 2 get on board. At the next stop, all stay on the streetcar, and 2 more get on. Next stop? Two leave the car, two more get on board. At the following stop, 10 climb on board, while 7 depart. At the next stop, 1 gets off; 5 get on. Question: How many stops did the streetcar make?"

The volunteer, and others as well, will protest that this is a "trick question," and they will be correct. Students had every reason to assume that the question would be, "How many people are now on the streetcar?" Nonetheless, the point is made that assumptions can be dangerous, as can jumping to conclusions.

In spite of the teacher asking an unexpected question, some students will be able to answer it correctly. To close the activity, the teacher may indeed want to ask students how many people were on the streetcar at the end (25).

Here is an entertaining variation of the Streetcar problem, sure to elicit both protests and laughs from students. Students may try to use their experience from the Streetcar problem to provide the correct answer here, but it will be to no avail (an individual student should be chosen to attempt to solve this problem, which should be presented to her orally):

"Pretend that you are a bus driver. The bus is traveling down Main Street with twenty passengers. At the first stop, half of the passengers depart, and four step on board. At the next stop, three passengers step off the bus, and the driver lets six more on. (The bus now makes a U-turn on Main Street, having reached the end of its run.) At the next stop, all but three passengers depart, and none board. Finally, at the next bus stop, seven passengers come on board. Question: what color is the bus driver's hair?"

Whatever color the hair of the chosen student is will be the answer. Remember, the first line of the problem was "Pretend that *you* are a bus driver."

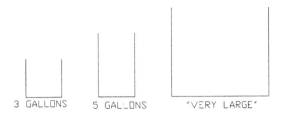

Figure 21.

#49: FOUR GALLONS, PLEASE

Curriculum Areas: Volume and Logic

Here is another classic problem that is likely to elicit plenty of shouts of "It can't done!" from skeptical and frustrated students.

"You are given two containers (see Figure 21). One holds exactly three gallons of water, the other, five gallons. There are no gradations on the sides of either container. You also have a very large container (we're not sure how much it holds, but it holds more than ten gallons). The large container likewise has no markings on its side.

Your job is to get exactly four gallons of water into the large container using only the equipment at hand. (You also have all the running water you may need from a nearby hose.) Can you do it?"

This is a good puzzle to assign as a warm up as students enter math class one day. The teacher may want to allow the student who solves it first to go to the chalkboard or overhead projector to provide an explanation of the solution.

#50: DOUBLE MY MONEY

Curriculum Areas: Doubling/Squaring

"The money in my pocket doubles every day, except on Sundays, when it triples. Today is Monday, June 1, and I have $0.10. When will I become a millionaire?"

This puzzler is most effective when the teacher insists upon hearing several student predictions—with rationale—before allowing them to use calculators to determine the answer. Unless students have had experience with this type of problem before, they will tend to overestimate the amount of time needed for $0.10 to become one million dollars. (Some will say several months; others, up to one year.)

The pocket will contain $1,415,577.60 on Tuesday, June 23.

The power of doubling (and tripling) is again illustrated by way of this problem. The terms "exponential growth" and "geometric growth" are two other ways of describing the phenomenon. The teacher may want to relate this challenge to Problem #63: What's a Mortgage?; both involve geometric growth.

Extension Activities

Students may complain that a pocket could not hold 1.4 million dollars. This would be a good time for the teacher to suggest that someone research American paper money denominations. (Is there a way for a pocket to contain over $1,000,000? If a student mentions that a check could be worth more than $1,000,000, the teacher has a great opportunity to review what checks are all about.)

It's not a bad thing for the teacher to insist that middle-grade youngsters memorize "the doubles" from 1 to 4,096. These numbers (1, 2, 4, 8, 16, 32, etc.) crop up often in the study of mathematics, and students will find that knowing them by heart will be especially helpful as they begin their study of algebra. (These "doubles" are, of course, powers of 2.)

#51: PALINDROMIC FUN

Curriculum Area: Sequencing

A palindrome is a word, phrase, sentence, or even a numeral that reads the same forward as backward. Single word palindromes include "mom," "radar," and "Bob." Sharing palindromes with students is an effective method for securing their attention at the opening of a class hour, and asking *them* to provide palindromic words or to invent palindromic sentences makes for an interesting optional assignment. The following is a look at both word and mathematical palindromes.

Here are some good word palindromes, preceded by some suggested explanations:

- A mini-biography of Teddy Roosevelt: "A man, a plan, a canal: Panama!"
- A middle school disciplinary tactic: "Draw pupil's lip upward."
- The first words ever spoken: "Madam in Eden, I'm Adam."
- Commentary regarding the comparative effectiveness of two cures for warts: "Straw? No, too stupid a fad. I put soot on warts."

Students should be warned that inventing palindromes of any length (and with even a tenuous connection to possible reality or to good grammar) is very difficult. For that reason many students will stay away from the challenge, just as some will accept it for the same reason.

Here are a couple of numerical palindromic curiosities:

- The squares of 1, 11, 111, and 1,111 (1, 121, 12,321, and 1,234,321) are palindromes.
- The square of 836 (698,896) is a palindrome. It is the smallest palindromic square containing an even number of digits.

CHAPTER 4

Demonstrations, Games, and Activities

#52: 7 × 13 = 28

Curriculum Areas: Arithmetic and Algorithms

THE 1940S COMEDY duo of Bud Abbott and Lou Costello once used the erroneous statement "7 × 13 = 28" as fodder for one of their routines. Believe it or not, several important math concepts can be reviewed with students by way of the teacher performing a solo version of the skit!

The teacher begins by simply writing "7 × 13 = 28" on the chalkboard, the overhead projector screen, or on the newsprint easel. When asked whether the times "fact" is true or false, most students will say "false." The teacher should "correct" those students by showing them the following steps of the multiplication algorithm (teacher "script" is given below as well).

Step 1. "Seven times three is twenty-one." (Teacher writes "21" as shown.)

$$\begin{array}{r} 13 \\ \times\ 7 \\ \hline 21 \end{array}$$

Step 2. "Seven times one is seven." (Teacher writes "7" as shown.)

$$\begin{array}{r} 13 \\ \times\ 7 \\ \hline 21 \\ 7 \\ \hline \end{array}$$

Step 3. "Twenty-one plus seven is 28, voila!" (as shown)

$$\begin{array}{r}13\\\times\ 7\\\hline 21\\7\\\hline 28\end{array}$$

As students continue to protest, the teacher should offer to "check" the answer—she should solicit from students that division is the check for multiplication. Here is how the teacher should then perform the division check by way of another "special" algorithm:

Step 1. "Seven into two won't go."

$$7\overline{)28}$$

Step 2. "Seven into eight goes once." (Teacher writes "1.")

$$7\overline{)28}^{1}$$

Step 3. "One times seven is seven." (Teacher writes "7" as shown.)

$$\begin{array}{r}1\\7\overline{)28}\\7\end{array}$$

Step 4. "Twenty-eight minus seven is twenty-one." (Teacher writes "21.")

$$\begin{array}{r}1\\7\overline{)28}\\7\\\hline 21\end{array}$$

Step 5. "Seven goes into twenty-one three times." (Teacher writes "3.")

$$\begin{array}{r}13\\7\overline{)28}\\7\\\hline 21\end{array}$$

"Proof," again, that $7 \times 13 = 28$!

$7 \times 13 = 28$

When the inevitable student protest reaches a fever pitch, the teacher should ask the students what the definition of multiplication is—students may or may not know to answer "repeated addition." If none provide the answer, the teacher should give it, volunteering to check her work by way of repeated addition, as follows:

Step 1. The teacher writes 13 seven times as a column:

$$\begin{array}{r} 13 \\ 13 \\ 13 \\ 13 \\ 13 \\ 13 \\ +13 \\ \hline \end{array}$$

Step 2. The teacher adds up the threes (three, six, nine, twelve, fifteen, eighteen, twenty-one). She writes the "21" at the top of the tens column:

$$\begin{array}{r} {}^{21} \\ 13 \\ 13 \\ 13 \\ 13 \\ 13 \\ 13 \\ +13 \\ \hline \end{array}$$

Step 3. The teacher adds all of the ones to the twenty-one (twenty-one, twenty-two, twenty-three, twenty-four, twenty-five, twenty-six, twenty-seven, twenty-eight!):

$$\begin{array}{r} {}^{21} \\ 13 \\ 13 \\ 13 \\ 13 \\ 13 \\ 13 \\ +13 \\ \hline 28 \end{array}$$

A good follow-up for this demonstration is for the teacher to have small groups of students deliberate for three minutes with the goal of identifying the teacher errors in each of the three algorithms. One student might then report findings aloud. This is another great trick that students should be encouraged to bring home to show their parents.

Teachers' Review of Algorithms

An algorithm is any established, step-at-a-time computational procedure for solving a problem. Setting up and solving the multiplication problem 8 x 23 in the following manner:

$$\begin{array}{r} \overset{2}{2}3 \\ \times 8 \\ \hline 184 \end{array}$$

is making use of an algorithm.

Algorithms provide useful, time-saving ways to solve problems. Yet they are not the only ways to solve problems. Students should not be given algorithms until they have had a chance to explore underlying math concepts first. (Students may be able to discover algorithms on their own.)

A danger in over-emphasizing algorithms in mathematics classes is that good student test scores may mask a profound lack of real conceptual understanding. Students who score well on tests of algorithmic skill may not in any way be able to apply those skills in real life.

#53: THE "ALL ANSWERS ARE NUMBERS" GAME

Curriculum Area: Connections

This simple game, while not involving counterintuition, holds an enormous amount of interest for adolescents. The many benefits of conducting this activity are discussed in the notes following the game directions.

The game: All students in the classroom stand at their desks or tables, or in a circle. The teacher then asks each student, in turn, to respond to the word or phrase given by the teacher with the appropriate number. For instance, the teacher might say, "Highest Monopoly® game dollar denomination," to which the student must answer "500."

To "digits in an area code" a student must reply "3." Students remain standing as long as they respond correctly. Those providing incorrect answers must be seated. The last student standing (after correctly answering a final query) is declared the winner.

This activity is a lot of fun, but there is a surreptitious motive here. Playing the game helps to provide students with reality-based connections for numbers. For instance, the number 52 may have absolutely no significance whatsoever to middle-grade students, until the teacher (or a student) points out that 52 is the number of playing cards in a standard poker deck. (Have they heard of "52 Pickup"?) If the teacher takes an additional moment with the youngsters to explore the four-suits/thirteen-cards-each relationship contained in every deck of poker cards, students will begin to establish a "feel" for the number 52 and the numerical relationships it contains.

Examine the "common" numbers 5, 10, and 12. If asked, students will likely be able to attach real-life significance to each. Five is the number of cents in a nickel, it's the age when most children first go to school, it's the number of Olympic rings. There are ten cents to a dime, ten years in a decade, and one should "count to ten" before getting angry. There are probably dozens of connections to the number 12, beginning with the number of signs in the Zodiac.

What about "uncommon" numbers such as 13 and 57? Students may or may not know about a bakers' dozen, and only some will know what the teacher is referring to when she connects the number 57 to the Heinz Company. The point is, the more such connections can be made for youngsters, the better. By way of this and related activities, students begin to see numbers as having significant characteristics beyond their positions on the number line.

A note about competition: As with any game that results in some students "winning" and some students "losing," the teacher has a serious responsibility to make certain that student self-esteem is preserved. Keeping things lighthearted, pointing out that many wrong answers may be good guesses nonetheless, and forbidding teasing are all ways in which the teacher can assure that positive learning occurs. The teacher is the role model regarding proper behavior during competitive gaming.

Extension Activities

(*1*) The teacher may allow students to provide the number clues. The instructor might let teams of students produce, say, 50 clues overnight for gaming the following day.

(2) The teacher can give students an unusual number, say, 714, and challenge them to attach some significance to it. (Babe Ruth held the record of 714 career home runs until it was broken by Hank Aaron in 1974.) The teacher can add interest to this activity by announcing that student answers will be rated as "good" or as "outstanding" as they are provided. For example:

The teacher provides the number 53. She might rate the response, "it's the next prime number after 47" as "good," while calling "It's the number of cards in a cheatin' deck" outstanding. (Identifying 53 as "the number after 52" should probably earn the student who offered that answer a good-natured round of boos.)

(3) The teacher should provide small groups of students with a series of whole numbers, say, 1–40. Each group should then be asked to privately label each number as "common" or as "uncommon." Groups can then report their ratings to the entire class. Do groups agree or disagree? Allowing students to defend their ratings would make for fine and perhaps heated discussion. ("Thirty-three is *not* a common number!" "Yes it is. It's an old vinyl recordplayer speed!")

Here are some starter clues. Teachers may want to make up their own lists of words and phrases—it wouldn't hurt to have several hundred at the ready.

Rings in a circus
Ships in Columbus' fleet
Leagues Under the Sea
Eyes on a Cyclops
Days in a leap year
Standard bubble-sheet-marking pencil number
Dots on a die
White House address
Day of Christmas (in the song) when calling birds are first given
Tally marks under a slash
Tentacles on an octopus
Blackbirds baked in the king's pie
Complete Van Gogh ears, late in life
Blind mice
Number of continents
First line of, "__, __, __, __, who do we appreciate?"
Your sweet teen birthday
Jefferson's coin

Faces on a cube
The number of the Interstate Highway which connects Detroit with
    Michigan's Upper Peninsula (or any local freeway reference)
Grams in a kilogram
Tea for *how many*?
Pennies in a roll
Buttons on a trumpet
Days in the longest months
Dalmatians
Number attached to PG movie rating
Bad luck Friday
Points for a touchdown
Dollars for passing "Go" in Monopoly®
Legs on a tripod
Scoops of raisins in Kellogg's Raisin Bran
Stripes on the US flag
James Bond's number
Hours in a day
Piano keys

Final Note: On his deathbed, an Indian mathematician is reported to have said that the number 1,729 is "captivating" because "it is the smallest number that can be expressed in two different ways as a sum of two cubes"!

#54: COMPLETE THE CLICHÉ

Curriculum Area: Connections

This well-known popular exercise is wonderful for warming-up students' brains at the beginning of math class. Students will at first plead ignorance when asked to identify the words that the letters below stand for, but once they get the hang of it they will greatly enjoy "completing the clichés," (many of which are not really cliches, but common phrases).

As with the preceding activity this one helps to establish the connection between numbers and reality. The teacher can call on individual students to attempt to find the words, or can assign the task to small groups as a low-risk competition. (Answers are provided.)

26 L of the A (26 Letters of the Alphabet)

13 S on the A F (13 Stars on the American Flag)
24 H in a D (24 Hours in a Day)
88 P K (88 Piano Keys)
57 H V (57 Heinz Varieties)
3 B M, S H T R! (3 Blind Mice, See How They Run!)
32 D F at which W F (32 Degrees Fahrenheit at Which Water Freezes)
9 P in the S S (9 Planets in the Solar System)
1000 W that a P is W (1000 Words That a Picture Is Worth)
54 C in a P D, with the J (54 Cards in a Poker Deck, with the Jokers)
3 R C (3 Ring Circus)
200 D for P G in M (200 Dollars for Passing "Go" in Monopoly®)
12 D of C (12 Days of Christmas)
20,000 L U the S (20,000 Leagues Under the Sea)
101 D (101 Dalmatians)
12 S of the Z (12 Signs of the Zodiac)
very difficult—The 4 H of the A (The Four Horsemen of the Apocalypse)

Naturally, students can be challenged to provide their own clichés or common phrases.

55: ARE YOU INFERRING . . .?

Curriculum Area: Inference

Presenting students with short statements, then asking them to make reasonable inferences from the statement, provides students with another good math class warm up. Asking them to provide unreasonable but possible inferences as well from those statements can make for an interesting and lively activity. Naturally, the teacher should be sure that students label unreasonable inferences as such. Here is one statement the teacher can post, along with a few of the kinds of inferences that the teacher should attempt to elicit from the students:

SHARKS

NO SWIMMING

If this were a sign posted somewhere, where might that be, and what might the sign mean?

The most obvious placement of the sign is at a public beach, and the intent is to keep people out of the water due to the presence of sharks. What are other possible inferences? (Remember, unreasonable but possible meanings are acceptable here.)

"The sign is posted at a school swimming pool, and the swim team known as 'The Sharks' is being notified that practice is canceled." (This is fairly reasonable.)

"The sign is posted at a public beach, and it is a warning to sharks that they are not allowed to swim in the area." (The fact that sharks can't read is irrelevant to the exercise.)

"The sign is a billboard announcing that, just as Michael Jordan *knows* basketball, and Babe Ruth *knows* baseball, sharks *know* swimming, and the word 'know' is misspelled."

A few other statements to post:

FIRE UP

THIS SPOT RESERVED

CHANGE

#56: AN INFINITY OF INFINITIES

Curriculum Areas: Number Lines and Infinity

This little demonstration makes for a good introduction to the study of infinity.

(Refer to the number line in Figure 22.) "If there are an infinite number of points between 0 and 1 on the number line, how many points are there between 0 and 2? *Double* infinity? If so, does that mean that some infinities are larger than others? And if *that* is so, does that mean that an infinity does not have to contain all of the elements of a class of things (such as points on a number line)?"

There exists an enormous (although not infinite) amount of information about the topic of infinity waiting to be researched by students. The importance of this demonstration is that it illustrates that infinity is

Figure 22.

not properly defined as "everything that there is" in a particular class of things.

This demonstration also helps to prepare students for their study of slope within the confines of algebra and trigonometry.

#57: YOU ARE MY DENSITY

Curriculum Areas: Volume and Density

This activity will help students better understand the concept of density, and become more accurate in their estimations of weight as well.

The teacher should fill a clear plastic trash bag with foam packing peanuts and tie it off. Any medium-to-large–sized bag will do—20 gallons is good. The teacher should also secure a balance-style scale (the large plastic type with two big trays on each side works best) and several small plastic and/or metal weights. Before class, the teacher should determine what combination of weights will balance the bag of packing peanuts, and he should keep that combination together in one pile. Ideally, the teacher will find a single metal or plastic weight that will perfectly balance the bag of peanuts. Naturally, the amount of peanuts can be adjusted until a balance is achieved.

With those items set up in advance, there are several good activities that the teacher can now try with his students to reinforce several density concepts and to promote meaningful discussion.

(*1*) After allowing several students to handle both the bag and the weight(s), the teacher should ask for predictions regarding which is heavier. Middle-grade students will typically state that the weight(s) is heavier. Once they put the items on the scale and see that they are of equal weight, the teacher should ask for comments and observations from students. Some will likely bring up, in one fashion or another, the concept of density (meaning weight per unit of volume). This will make for a lively discussion. Look for students to use words such as thick, thin, dense, fluffy, solid, airy, and foamy.

(2) A variation of the above activity is to challenge students to make the weight of the bag of packing peanuts equal to a designated weight in advance of using the scale. Students can work in pairs, small groups, or individually.

(3) The teacher can turn this activity into an appropriate classroom contest. For instance, a small prize can be awarded to the student (or duo, or small group) who can best match the weight of the bag to the given metal or plastic weight in advance. (The teacher should remember to allow students to do the actual weighings.)

Extension Activity

Students might be interested in knowing how much all of the air in the classroom weighs. Finding this out, and reporting back to the class, makes for a good optional extension challenge for one or a team of students. The report will reinforce the idea that when weight is distributed over a large volume, the sense of weight tends to disappear. (For the teacher: a cubic foot of air at sea level and at room temperature weighs about 1 ounce.)

#58: THE SPREADING VIRUS (AKA THE SPREADING SECRETS)

Curriculum Area: Probability

The following is an unusually powerful and effective activity. It requires a good deal of preparation on the part of the teacher, but it is well worth the hassle. Its roots are in mathematics, but its theme is sociological.

The Spreading Virus activity can be used to demonstrate how quickly the HIV virus, for instance, can be spread to many people, even if the infected persons do not have numerous sexual partners. Deciding whether or not to use the Spreading Virus activity to make that point is a decision that the teacher should make in consultation with the school principal. (If it is decided that referring to HIV is inappropriate, the activity can also be used to demonstrate how the common cold is quickly transmitted. For even younger students, "passing secrets" can be the theme.)

The teacher will need a supply of two chemicals: sodium hydroxide and phenolphthalein. Many high school or even middle school science laboratories will have these common chemicals. Both are colorless and odorless—they look like water. Obviously, the teacher should take great care in the handling of these chemicals. The teacher will also need about ten very small clear plastic cups, and a supply of water.

In advance of the activity, the teacher should half-fill nine of the cups with plain water. A tenth cup should be half-filled with sodium hydroxide. The teacher will need to keep careful track of the location of the cup containing the sodium hydroxide—all ten cups will look the same. These ten cups should be placed on a tray and kept in a safe spot until the beginning of the activity.

At the start of the activity, the teacher should announce that ten students will be chosen to randomly select one of the ten cups each; others in the classroom will be spectators. (The teacher should secretly note which youngster has taken the cup containing the chemical.) Once the ten students have chosen their cups, the teacher should give the following instructions:

(*1*) At her word, each student should "mingle" with one of the other ten (randomly picked) by carefully mixing the contents of their cups back and forth several times. If the teacher has chosen to use this activity as a demonstration of HIV spread, she should announce that each "mingling" represents a single sexual contact. Other options are for the teacher to state that each "mingling" represents someone sneezing in someone else's face or the telling of a secret to "just one other person."

(*2*) Students should then "mingle" again with someone new; the mingling again representing a sexual contact, a sneeze, or the passing of a secret.

(*3*) Finally, all students should "mingle" for a third time, again, with someone new.

Experienced middle-grade teachers will know to expect a good deal of laughing and joking from students if the HIV option has been chosen, in spite of the seriousness of the topic. This is normal, but the teacher will want to "keep a lid on" things.

Next, the teacher should produce the phenolphthalein, along with an "eye dropper" meant for chemical use. Phenolphthalein is a sodium hydroxide indicator—a drop added to sodium hydroxide will immediately turn it purple. Even a weak water/sodium hydroxide solution will change color.

The teacher should then put a single drop of phenolphthalein into each of the students' cups, revealing who has become "infected." (A minimum of four solutions will turn purple; more than four is likely.)

A spirited discussion is now sure to ensue. A goal of the discussion—without teacher prompting needed—will be for the group to discover who the original "infected" individual was. This can always be done if students approach things in a logical manner.

These questions may be thrown out to students as well:

(*1*) What might have been the outcome if everyone had had only two "partners"?
(*2*) How many "minglings" do you think would have been needed to *guarantee* that all ten participants were "infected"?

It is left to the teacher to see to it that students understand the point regarding the inadvisability of ignorant sexual contact illustrated by this activity.

#59: HOW'D HE DO THAT? PART I. THE KANGAROOS-IN-DENMARK TRICK

Curriculum Area: Mental Arithmetic

After soliciting a volunteer, the teacher should ask her to perform the following computations and other tasks mentally. For this trick, it's best if the teacher chooses a youngster who is good at mental arithmetic. (The rest of the class should be welcomed to follow along in their heads, but should be cautioned against saying anything aloud.)

(*1*) Mentally choose a whole number less than 10.
(*2*) Multiply that number by 9.
(*3*) Add together the digits of that product.
(*4*) Subtract 5 from that sum.
(*5*) Take that number and go to its corresponding letter of the alphabet (A = 1, B = 2, C = 3, etc.).
(*6*) Think of a country whose name begins with that letter.
(*7*) Take the last letter of the name of the country you've chosen, and think of an animal whose name begins with that letter.

The teacher should now ask the student to "think intensely" about the country and the animal chosen, and the teacher should indicate that she is trying hard to "receive" those two names. A furrowed brow on the teacher's forehead would be effective at this point.

After a moment of this intense thought, the teacher's expression should shift to one of surprise and shock, and she should loudly announce that, "You can't use those. *There are no kangaroos in Denmark!*"

The teacher's announcement will absolutely shock some students if the trick is done well. For a few moments, some will be completely convinced that the teacher is able to read the mind of the volunteer. Other students who followed along in their minds will also be momentarily stunned. All of this will quickly dissipate as students realize that everyone in the class who chose to think along with the volunteer has picked Denmark and kangaroos! Here's what's at work in this trick (teachers should remember to allow students to try to come up with the explanation).

Regardless of which number the volunteer originally chooses, she will be using a four before beginning step five. The magic of the number nine is casting its spell: The sum of the digits of a two-digit multiple of nine is always nine. When the volunteer is asked to subtract five from the nine, the inevitable result is four.

When asked to go to the corresponding letter of the alphabet, the student finds herself at the letter d (still thinking, perhaps subconsciously, that she could very well be at any of the other letters of the alphabet). Here is where the teacher is playing the odds: The first country that comes to mind *for most people* when asked to think of one beginning with a d is Denmark. And what animal comes to most minds when the first letter is k? A kangaroo, of course!

In addition to having some fun with the students, this activity affords the teacher the opportunity to review some special characteristics of the number nine and of its multiples.

#60: HOW'D HE DO THAT? PART II. THE 20-TO-50 TRICK

Curriculum Area: Mental Arithmetic

Students will delight in having been fooled by the teacher with this classic trick. As always, the teacher can promote quality discussion and even debate by refusing to provide an early explanation of how the trick was done—that's the students' job! Here's how the 20-to-50 Trick appears to the youngsters.

The teacher asks for a student volunteer (without, by the way, explaining the nature of the task to the student in advance). The volunteer is asked to stand (which will help to further promote an appropriate air of nervous anticipation). The teacher asks the student to name aloud a whole number which meets the following simple criteria:

(*1*) The number is between twenty and fifty.
(*2*) Both digits are odd.
(*3*) The two digits are *not* the same.

The moment the student has announced his choice (say 37), the teacher will without hesitation pull out of his pocket a folded slip of paper, which is then handed to the volunteer, who is invited to unfold it and to show it to the rest of the class. The number boldly written on the slip is . . . 37!

With this trick, all that the teacher need do as a follow-up is fold his arms, smile and nod at students as they together determine the secret of the trick. How does the teacher pull off this one? Here is the set-up.

Before class, the teacher needs to write the numbers 31, 35, 37, and 39 on separate slips of paper. Why only those numbers? Because no other whole numbers fit the given criteria! Then, the teacher should put one slip in each of his four pants pockets. If the teacher does not have four pants pockets, shirt pockets will do, as will cuffs, or even shoes. When the student volunteer announces her number choice, the teacher simply needs to recall where the particular corresponding slip is located, and to "magically" produce it, to the amazement of one and all.

This activity provides a good surreptitious review of the definitions of *digit* and of *odd*. It also points out how quickly providing just a few qualifiers can drastically reduce the number of numbers still under consideration. Students are given the "feel" at the beginning of this problem that they will be randomly choosing from a wide range of numbers, not sensing after having been given the restrictions that they in reality have few to choose from.

The end of the class period is the best time to do this trick—waiting until then helps to reduce any sense of a "set-up."

Encouraging students to try this trick on their parents and siblings is a great way to promote the school/family connection.

CHAPTER 5

Investigations

#61: ZENO'S PARADOXES

Curriculum Areas: Geometry and Infinity

THE FOLLOWING IS not actually a counterintuitive *problem*; rather, it's a retelling of a classic paradox (a statement containing two apparent truths that are mutually exclusive and contradictory) first offered by the fifth-century Greek philosopher Zeno of Elea. Many middle school students are intrigued by concepts such as infinity, and this activity provides a great springboard for discussion. Some of the deeper thinkers in math classes may find the implications of this paradox downright mind-boggling.

A nice feature of this paradox is that it can be acted out in class by the more kinesthetically inclined students.

The Paradox

Movement is impossible, and here's why: Let's say that someone would like to walk from point A to point B (the front of the classroom to the back, perhaps). In order to get from A to B, the walker must, of course, pass the mid-point of segment AB. (Here's where it gets interesting.) Once the walker reaches that mid-point, she has a finite distance remaining in her walk (half the original distance). She must, of course, pass the halfway point of the remaining distance, and she will continue to have to pass new, evermore tightly spaced halfway points as she nears the far wall, over and over again. The difficulty is this: There are an infinite number of halfway points for her to pass, and she needs to pass them one at a time. Given that there will always be another halfway point to pass, she can never reach the wall.

The Paradox, of course, is that she *can* reach the wall, in spite of the impeccable logic that says she can't.

The teacher will want to challenge students to find an error in Zeno's logic. Most any student attempt at providing thoughtful analysis should be accepted, at least at first. The key to sustaining good discussion is for the teacher to continue to solicit student reaction to classmates' remarks.

Extension Activities

(*1*) Some students may be motivated to research Zeno and his other paradoxes (see below), and to make brief presentations to the class. Teachers should encourage those students to include physical demonstrations of other paradoxes when reporting.

(*2*) Ask students to provide examples of "real-life" paradoxes, verbally or in writing. Here are two:
- "We only hire workers with experience."/"You can't get experience unless someone hires you."
- Said to a teenager: "Will you please act grown-up."/"You can't do that, you're just a kid."

This can be done as a classroom brainstorming activity, or, as an individual or small group project.

A Note on the So-Called Bumblebee Paradox

A statement presented as a paradox has circulated in America, and perhaps elsewhere, for many years. It provides leaders of "positive-thinking" seminars with a great opening line; the problem is, it's not a paradox at all, it's just a false statement. It goes like this: "Scientists have proven conclusively that, according to the principles of aerodynamics, bumblebees cannot fly. Their wings simply cannot possibly support their weight. Bumblebees, however, know nothing of the principles of aerodynamics; they fly just by doing it." The truth is that the aerodynamics of insect flight are well understood by scientists; it's doubtful that any scientist ever made such a ludicrous statement.

Another of Zeno's Paradoxes

This paradox also provides a good opportunity for students to illustrate by way of acting out.

Achilles can run ten times as fast as a tortoise, and the two are pitted against one another in a race. The tortoise is given a head start of 10 units. Zeno states that Achilles will never pass the tortoise. Here's why. Each time that Achilles reaches the point where the tortoise was, the tortoise will have advanced one-tenth of the distance that Achilles has run. Thus, the tortoise will always be in front.

#62: HOW ABOUT A NICE PIECE OF PI?

Curriculum Areas: Circles and Pi

The number which represents the ratio of the circumference of any circle to its diameter is designated by the Greek letter pi (π). The ratio has a fascinating history, and it affords many opportunities for enrichment and student engagement in the middle school mathematics classroom.

Historical Background

Many of the relationships contained within simple geometric figures exude a certain simplicity. For example:
- Multiplying the length of one side of an equilateral triangle by three will give the perimeter.
- A square has exactly four equal-length sides.
- The ratio of the length of a square's diagonal to its perimeter is 2:4.

Relationships within circles, however, may appear to some as less than elegant. In fact, it is impossible to completely and accurately note the relationship known as pi as a number in decimal notation. Mathematicians have devised clever ways of allowing us to accurately note the numerical representation of pi to as many digits as we care to know, but as a decimal, pi is both non-terminating and non-repeating.

I Kings 7:23: "Now he made the pool of cast metal ten cubits from brim to brim . . . and thirty cubits in circumference." This biblical reference to the work of Hiram the Builder shows that even the ancients knew that pi was *about* 3. Later, the Greek mathematician Archimedes correctly stated that pi was between 3-10/71 and 3-1/7, or, in decimal form, between approximately 3.14084 and 3.14286. (He used actual polygons of 96 sides to obtain his figure!)

The ancient Chinese astronomer Ch'ung Chih gave pi as being between equal to 355/113, or 3.141593, an extremely accurate estimate.

Gottfried von Leibnitz, a 17th-century German mathematician, used calculus to express pi as an infinite sum:

$$= 4\left(1 - \frac{1}{3} + \frac{1}{5} - \frac{1}{7} + \frac{1}{9} - \frac{1}{11} \ldots\right)$$

Other modern methods of calculating pi (dependent upon the advent of computers) allow us to calculate pi to the nth decimal place.

Activities

Students will gain a good basic understanding of the meaning of pi if they are allowed to measure the diameters and circumferences of many circles. They can then compare the lengths of diameters and circumferences, first by trying to guess the ratio ("close to three" will be common), then by way of division on the calculator, discovering that pi is a bit larger than 3.

The teacher must make certain that, at some point in this preliminary investigation, the point is made that the ratio known as pi is consistent for all circles.

It would be helpful if at least the first one hundred digits of pi as a decimal were posted in the classroom. (Several mathematics supply catalogs offer such a poster; it's generally about eight inches by twenty feet in size. Some companies offer a poster which shows the first one thousand digits.)

A key theme when discussing pi is that there is no pattern to the sequence of digits. The teacher may want to challenge students to try to find patterns in the first one hundred or one thousand digits, and some may insist that they indeed have. (For instance, within the first two hundred digits there is a sequence of three consecutive 5s.) Here's what's at work—this should be shared with youngsters.

There are only so many ways to arrange the ten digits of our base ten system of numeration, and eventually, with a long enough string of digits, what appear to be meaningful patterns will appear. This is absolutely inevitable. (A list of the first one thousand whole numbers, 1–1000, contains the numbers 111, 222, 333, etc., yet this is not surprising, it's natural.) If enough events take place, some will appear "unlikely." The odds of a particular youngster being exactly who she or he is, are incredibly small, yet all of us are indeed exactly who we are!

Another profitable exercise is for the teacher to ask students to tally the number of ones, twos, threes, etc., in the first one hundred digits of pi (or better yet, the first one thousand). What they will find is that each of our system's ten digits appear about 10% of the time! (The teacher should, of course, solicit student predictions before allowing them to take the tally.)

Middle-grade students will undoubtedly be impressed by the fact that at least one person in the history of the world memorized pi to 10,000 places! (The teacher may want to query students as to how that person might have accomplished the dubious feat. The teacher should give students credit if they can connect this man's feat to problem #64: A Few Notes . . .

A clever mnemonic device for the first few digits of pi is "Yes, I know a digit." and "May I have a large container of coffee?" The number of letters in each word gives the first few digits of pi. Students can be challenged to develop other mnemonics for pi.

Allowing further research into the history of pi makes for excellent student assignments.

For convenience and reference, here is pi as a decimal, taken to one hundred places:

3.14159 26535 89793 23846 26433 83279 50288 41971 69399 37510
58209 74944 59230 78164 06286 20899 86280 34825 34211 70679

#63: WHAT'S A MORTGAGE? (OR, "I HAVE TO PAY BACK HOW MUCH?")

Curriculum Areas: Consumer Mathematics, Percent, and Compounding

Middle-grade students may think as the author once did: He thought that to pay back a home loan of, say, $60,000 at 8% interest, he would eventually have to return $60,000 plus 8% of $60,000, a total payback of only $64,800! Little did he know, as students may not, that the 8% was an annual rate applied to the outstanding balance every year. The actual payback for such a loan over twenty years is over $180,000.

"Mortgage" is simply a term for a loan given for the purchase of a house or other real estate. Sharing this definition with students should take some of the mystery out of a term they have likely often heard, but never quite understood. (As always, the teacher should give the students a chance to provide definitions of "mortgage" before offering her own definitive version.)

A look at mortgages with students affords the teacher the opportunity to reinforce earlier discussions of compound interest—this, of course, is what is really at work here. A good calculator activity is this: Supply students with an original principal amount (say $80,000) and an annual interest rate (say 9%). Allow students to see how the amount owed increases rapidly year-to-year if no payments are made (80,000 × 1.08 for year one, 86,400 × 1.08 for year two, and so on). In real life, one cannot go for years without making payments, but the exercise illustrates again the power of compounding.

A somewhat more true-to-life exercise would have students "making payments" each year, then compounding on the remaining balance. The most realistic activity would have youngsters "making payments" each month, and compounding interest each month as well. This is not so much a difficult activity as one that students and the teacher may find tedious. Even with a calculator, the need to complete the long series of multiplication, addition, and subtraction may diminish the value of the activity. It may be worth a try nonetheless.

When looking at mortgages, it might behoove the teacher to invite in an outside speaker; a banker, perhaps.

#64: A FEW NOTES . . .

Curriculum Area: Estimation

As part of this investigation, the teacher will need to play a recording of a pianist performing an interesting solo piece. (An attractive alternative can be for a student, the teacher, a guest teacher, or a visitor to play the piece.) The selection can be short (two–three minutes) or much longer (10–20 minutes), depending upon how much time the teacher would like to devote to this activity.

The discussion which follows is based upon the teacher playing a piano solo version of Gershwin's *Rhapsody in Blue*. Other suggestions are noted below.

Although it is a semi-classical composition, *Rhapsody* is suggested because it is an unusual piece, full of clever melodic twists and turns that might engage the adolescent. Students may recognize one of the themes in *Rhapsody* as being used in United Airlines television commercials (as usual, the teacher should not tell this to students; he should let *them* make note of this).

Original recordings of *Rhapsody in Blue* played by George Gershwin in the 1930s are readily available. Using one may open some interdisciplinary doors (the Jazz Age, the Great Depression, etc.).

The Problem

"How many individual musical notes do you think the pianist produces when playing this piano solo version of *Rhapsody in Blue*?" (Teacher or performer then plays the piece, or the recording is played.)

Students will tend to be fairly wild in their guessing, with underestimation being the norm. As they wrestle with methods for estimation, the teacher may want to provide this suggestion: "When a nurse takes your pulse, does she listen to your heartbeat for a full minute? ('No; she listens for 15 seconds, then multiplies by four.') How might that fact help you estimate the number of notes played in *Rhapsody in Blue*?"

If the performance was live, with someone actually at the piano, students should be encouraged to use the performer as a partner in experimentation. They may ask the pianist to play what is usually 30 seconds of the piece as slowly as possible, attempting to count each note, with the intention of multiplying to come up with a good estimate of the total. One student might count right-hand notes, the other, left.

If the sheet music is available, there will be at least one youngster willing to actually count the notes, one by one, as an out-of-school project. This should be encouraged. A next-day comparison with previous estimates should be fun.

Extension Activity

A good follow-up to "A Few Notes . . ." is a teacher-led discussion of combinations and memorization.

Consider: If Gershwin played his Rhapsody without sheet music in front of him (which was always the case), that means that he had to memorize every single note, exactly when to play it, and what pressure to apply to the keyboard to produce each sound. Furthermore, he had to correctly play each note as a whole, half, quarter, eighth, dotted quarter, etc.—there are perhaps twenty choices here per note. Given that most people cannot remember the ten digits of a phone number for more than a few seconds, how is Gershwin's (or any good pianist's) playing possible? How can they play thousands of correct notes with all of their characteristics exactly right within minutes?

The response that "He was a genius," will not suffice. Many merely *good* piano players and other musicians are capable of performing nearly the same feat. What's at work here?

Here's an activity for middle schoolers which will help to bring light to that mystery: The teacher should boldly display the following cryptic sequence on the chalkboard, the overhead projector, or on the computer display screen:

N   BCFB   ITV   TG   IFN   BA

Students should be told that they have 30 seconds to memorize the sequence—no writing it down!—and that they will then be asked to correctly repeat it aloud (with the sequence hidden, of course). Students will be quite enthusiastic about this challenge. Few will be able to do it.

Next, the teacher should show this sequence, again asking students to memorize it in 30 seconds:

NBC   FBI   TV   TGIF   NBA

It's likely that, even before the 30 seconds are up, someone will notice that this second sequence is the same as the first. Most everyone will also note that the second version is easy to memorize.

The teacher should ask the students why the second sequence was so easily memorized. When it is clear to all that "organizing the letters into recognizable and not random chunks" is the key, the teacher should ask students to relate this exercise to the music counting-notes challenge. They will likely be able to state (in their own words) that organizing information into meaningful units makes managing the information much easier.

Did Gershwin (or does any musician) memorize the characteristics of each note, then refer to those characteristics when performing? Hardly. Rather, notes are organized into chords, groups of chords into bars and measures, and so on, with each chord, or bar or measure having a distinct "flavor," which the musician recalls and then reproduces (just as the acronyms NBC, FBI, and TV have their own "flavor").

A note on combinations: Of all of the possible combinations of notes which composers can assemble, only a very few combinations produce "music." A musical staff can note, in order, b, b-flat, d, a, a-flat, and g, but that sequence may not qualify as music. Given a thousand notes, perhaps less than one-tenth of one percent of all possible combinations will produce a tune.

An important point is contained herein. Insisting that students memorize mathematical rules and formulas without allowing them to have meaning attached to the rules and formulas is folly. Teachers should beware, too, of good student grades on rule/formula/algorithm-oriented tests. Those good grades may simply be an indication that the student is able to commit things to his or her short-term memory, just as most of us can memorize a new phone number just long enough to dial it.

Here are a few other suggestions for musical pieces:

- *The Flight of the Bumblebee* (violin), approximately two minutes
- *Moonlight Sonata* (piano), approximately six minutes of slowly played music
- *Hungarian Rhapsody* (piano), approximately ten minutes of very lively music

The teacher should be sure to inform the music teacher about this lesson. The interdisciplinary approach will be appreciated.

A final note: The solo piano version of *Rhapsody in Blue* contains over 20,000 notes.

#65: BOOTS IN HOBBLE JUNCTION

Curriculum Areas: Percent and Logic

This is a wonderful problem with a double twist, as will be seen shortly.

"In the old western town of Hobble Junction lived 1,000 people. Exactly 10% of the residents were one-legged (from all of the shoot-outs) and half of the rest of the people in town always went barefoot. Assuming that everyone except the barefooters wore boots, how many boots were there in town? (Assume, too, that no one had any extra pairs of boots.)"

These will likely be the steps taken by most students in attempting to solve the problem:

(*1*) 10% of 1,000 is 100, so there we have 100 one-legged people, or 100 boots.
(*2*) There are 900 people remaining. Half of 900 is 450, and those 450 people each wore two boots; that's another 900 boots.
(*3*) 100 + 900 = 1,000, so there were 1,000 boots in Hobble Junction.

Students who give 1,000 boots as the answer are correct, and that should be recognized by the teacher. There is an interesting twist to this problem, however, and once the teacher has praised the students for solving it, she should give them the following slightly different version of the Hobble Junction boot problem

"In the old western town of Hobble Junction lived 1,000 people. Exactly 20% of the residents were one-legged (from all of the shoot-outs) and half of the rest of the people in town went barefoot. Assuming that everyone except the barefooters wore boots, how many boots were there in town? (Assume, too, that no one had any extra pairs of boots.)"

(Note that the only variation from the first problem is that now 20 and not 10% of the population is one-legged.)

Students may at first question why so similar a problem is being presented, but the teacher should insist that they solve it. They will likely do so very quickly, and the teacher will undoubtedly enjoy watching students' faces as they realize that they have again come up with an answer of 1,000 boots!

Hopefully, students will now pose this question on their own: "If the amount of one-legged people is changed to any given percentage, will there always be 1,000 boots?" The teacher should allow students the chance to check out this proposition (yes, the answer will always be 1,000 boots) and try to explain the phenomenon. If students are unable, after guided discussion, to state why the answer will always be 1,000, the teacher may want to supply them with the following explanation.

Forget for a moment that a certain portion of the population wears exactly one boot. The fact is that, for the remaining population, *there will always be as many boots as there are people.* Half of that group wear two boots, the other half wear none; the average is always one boot per person. Add one boot per one-legged person, and you have 1,000 boots, always!

#66: ESP?

Curriculum Area: Probability

The five drawings shown in Figure 23 are universally considered to be "ESP" symbols. They are used in experiments that purport to reveal a person's extrasensory abilities. The following activity allows the teacher and her class to explore probability theory, while inculcating a proper sense of skepticism into students regarding claims of ESP (although there is no intention here to rule out the possibility of the existence of the phenomenon).

Two important notes:

(*1*) The teacher should not allow this activity to turn into an attempt to discover anyone's supposed ESP. *Probability* is the main topic.

(2) Some parents might find ESP experiments objectionable—another reason for the teacher to emphasize only the mathematics involved.

Typical experiments with these symbols involve two individuals; one who randomly selects one of the symbols and "thinks hard" about it; another who attempts to "see" which symbol the first is thinking

Figure 23.

about. Identifying the correct symbol often enough is seen as evidence of ESP abilities.

As an introduction to the mathematics involved with such experiments, the teacher should ask the students to imagine that perhaps one hundred individuals are being tested, one at a time and separately, in the following manner: Each subject is asked to name which randomly chosen symbol the experimenter is "thinking about" five different times. She should then lead a class discussion that includes students' responses to the following questions:

(*1*) What do you make of a subject who correctly guesses two out of five symbols?
(*2*) What do you make of a subject who correctly guesses three?
(*3*) Let's say that person after person does no better than guessing three out of five symbols correctly, but that the 92nd person to be tested correctly identifies five out of five. What would you say that this proves? (Students may be split on this issue. Some will say that five out of five indicates ESP, others will note that eventually, guessing alone will result in someone getting five right.)
(*4*) If we are skeptical that the 92nd subject has ESP, what might be a way to lessen our skepticism? (Students will likely state that the experiment would need to be repeated several times.)

At this point, the teacher may want to inject a review of the Raffle problem, #7, into the discussion. When one thousand tickets are sold for a raffle, and only one winning ticket is to be drawn, someone is going to win in spite of the fact that the odds were 1,000-to-1 against him or her! The connection with the ESP discussion may be obvious to middle-grade students:

Eventually, the unlikely is nearly certain to happen.

In a sense, all events are nearly infinitely unlikely, but they happen. Let's say that the next car that passes on the street in front of the school is a red 1998 Plymouth Breeze with a small dent in the right front fender and with an orange tennis ball affixed to the antenna. What were the odds that a car matching that description would pass by at exactly that moment? Nearly infinitely small; no better than one in a billion, at best. Yet, these nearly impossible events make up all of the absolutely routine events of our days!

The next investigation, Roll the Dice, provides a look at probability from another angle.

#67: ROLL THE DICE—A PROBABILITY INVESTIGATION

Curriculum Areas: Probability, Fractions, Decimals, and Tables

"What are the possible outcomes when rolling two regular six-sided dice?" Most middle-grade students are able to state that the possibilities are the sums two through twelve. But when asked, "Are each of those eleven possible outcomes equally likely to happen every time the dice are rolled?" many students will incorrectly say "Yes." This is due to student confusion over the difference between the terms *possible* and *likely*.

Here is a kinesthetically oriented activity that will help students better understand the probabilities inherent in the rolling of dice. An entire class hour or more will be needed for this activity. The teacher will need to supply the class with as many dice as there are pairs of students in the class—they will be working in pairs. The specific question this activity will answer for students is "Which dice rolls come up most often, which come up least often, and why?" That is, students will *quantify* outcome probabilities.

Before attempting this activity, the teacher should see to it that most students have a good understanding of decimal notation, and the ability to convert uncommon fractions to decimals with a calculator. Students also need to know that odds normally expressed as fractions can also be expressed as equivalent decimals.

For example: Students know that the odds of a flipped quarter landing with heads up are "50/50," or "50 times out of a hundred" for both heads and tails, or, "1/2" for heads and "1/2" for tails. The teacher should make sure that students see that this can also be expressed as odds of "0.5" for both heads and tails. The teacher may want to check for student understanding by asking them to convert these fractional odds to decimals: "three out of four," "2:3," and "13/100."

The teacher should explain that the class is going to make several hundred dice rolls, keeping careful track of outcome frequencies. Students will work in pairs, with class tallies later being added together for analysis.

108 INVESTIGATIONS

The teacher should give each student duo two dice after providing them with these instructions:

(*1*) Designate one partner as a "roller" and the other as a "tallier." (Students can switch roles mid-activity if they wish.)

(*2*) The roller should toss the pair of dice 100 times, with the tallier keeping track of outcomes. (The tallier should also keep a running tally of how many times the dice have been rolled—the roller should not be allowed to roll past 100 rolls.) A completed group tally sheet should look something like the sample below.

Number of Times Each Possible Outcome Occurs for My Group of Two Students

2	|||	3
3	‖‖‖ |	6
4	‖‖‖ |||	8
5	‖‖‖ ‖‖‖	10
6	‖‖‖ ‖‖‖ |||	13
7	‖‖‖ ‖‖‖ ‖‖‖ ||||	19
8	‖‖‖ ‖‖‖ ‖‖‖	15
9	‖‖‖ ‖‖‖ |	11
10	‖‖‖ ||	7
11	‖‖‖ |	6
12	||	2

When it appears that all pairs are done with their task, or nearly so, the teacher should ask for a report from each group. (*Important note:* In order for a group to report, they should turn in their dice to the teacher first!) That report will take the form of the tallier announcing aloud to the teacher how many twos were rolled, how many threes, etc., up to how many twelves were rolled. The teacher should keep track of each group's results on a prepared master grid on the overhead projector or on the chalkboard. Below is a sample master grid.

Number of Times Each Possible Outcome for the Entire Class, Shown as a Fraction (out of 800), Then as a Decimal

2	3	4	2	2	3	0	5	3	27/800	0.028
3	6	4	6	4	6	8	5	7	46/800	0.058
4	8	9	9	10	8	5	9	9	67/800	0.084
5	10	10	9	11	10	9	12	9	80/800	0.100
6	13	12	14	13	13	14	10	14	104/800	0.130
7	19	20	19	21	19	18	19	19	155/800	0.194
8	15	14	15	13	15	16	14	15	117/800	0.146
9	11	12	11	11	10	11	9	11	86/800	0.108
10	7	5	6	7	8	9	10	6	58/800	0.073
11	6	7	7	7	6	5	7	7	52/800	0.065
12	2	3	2	1	2	5	3	0	18/800	0.023

Students will note with interest how similar each groups' report is—few twos and twelves, lots of sixes, sevens, and eights.

When the master grid is complete, the teacher should solicit student help in finding the number of total rolls for the eleven possible outcomes (2–12). Those totals should be expressed first as fractions (800 being the denominator for all outcomes in the sample master grid provided—this may vary classroom-to-classroom). With the help of students and their calculators, those fractions should then be converted to decimals. The decimals should be rounded to the nearest hundredth place, as shown in the sample master grid.

The teacher will now want to ask students to state the significance of the various decimals. "What does the 0.028 for the twos tell us, as opposed to the 0.194 for the sevens?" Students will see that the closer an outcome is to seven, the more frequently it is rolled, with seven itself being the most frequently rolled outcome.

This follow-up activity (perhaps, by necessity, done the next day) will help explain the "why?" of the results to students.

The teacher should ask students this question: "What are all of the ways to get a total of two when rolling the two dice?" The answer, of course, is by rolling two ones—there is no other possibility. The same question should be asked for all of the possible outcomes with two dice: "What are the ways to roll a total of three? four?" and so on through twelve. The teacher or a student should again keep track of student responses on the overhead projector or chalkboard, producing a chart similar to the one below.

Possible Ways of Rolling the Outcomes 2–12 with Two Dice, Shown as a Fraction (out of 36), Then as a Decimal

2	(1,1)	1/36	0.028
3	(1,2)(2,1)	2/36	0.056
4	(1,3)(3,1)(2,2)	3/36	0.083
5	(1,4)(4,1)(2,3)(3,2)	4/36	0.111
6	(1,5)(5,1)(2,4)(4,2)(3,3)	5/36	0.139
7	(1,6)(6,1)(2,5)(5,2)(3,4)(4,3)	6/36	0.167
8	(2,6)(6,2)(3,5)(5,3)(4,4)	5/36	0.139
9	(3,6)(6,3)(4,5)(5,4)	4/36	0.111
10	(4,6)(6,4)(5,5)	3/36	0.083
11	(5,6)(6,5)	2/36	0.056
12	(6,6)	1/36	0.028

When the students have helped the teacher complete the left side of this "possibilities chart," he should ask students to count up the total number of different two-dice possibilities displayed on the chart—they should say "36." Next, the teacher should ask students to count how many of those 36 possible outcomes "belong" to each total 2–12. (There is one way to roll a two, two ways to roll a three, three ways to roll a four, and so on.) These should be displayed on the grid as fractions (see the chart above), then those fractions should be converted by calculator to decimals (nearest hundredth) and noted on the chart.

If students are observant, they will see that the decimal probabilities on each of the two charts are nearly the same. The teacher should now reinforce that these decimal probabilities are reliable, meaning, for instance, that students can refer to them when playing certain games. (If a youngsters' marker is two spaces away from the highly desirable Boardwalk property in the game of Monopoly®, she should not count on rolling a total of two. The odds are very much against it!)

Students will find this activity enjoyable if only because they are allowed to make what most adults find to be highly annoying noise (the rolling of the dice) for many minutes!

APPENDIX A

Problem-Solving Strategies

MIDDLE-GRADE MATHEMATICS classrooms ought to be hotbeds of problem-solving activities. Teachers can provide the activities (this book is intended to help in that regard); they can also teach students specific problem-solving strategies. This appendix provides a brief review of some of those strategies. First, lets look at the terms problem and problem solving.

A *problem* is a task or challenge or situation that requires one to develop a solution, even though the methodology may not be readily apparent. Problems should not be seen simply as computational tasks, not even multi-step ones, although computational tasks are certainly problems as well.

Problem solving is "figuring out" a problem. It may involve finding a specific numerical value; it may mean spotting a pattern, or it may mean being able to explain how the "difficulty" may be resolved.

GOOD PROBLEM-SOLVING TECHNIQUES

(*1*) *Guess and check.* This is an exceptionally valid technique for students to use; it is not to be confused with "wildly and randomly guessing." Most people's guesses include a good deal of inherent wisdom and insight, even when that insight cannot be articulated. Here is a problem which can be solved algebraically, but which many students will intuitively solve using the guess and check method.

"Three persons' ages add up to 186. Their ages are consecutive even whole numbers. What are their ages?"

Typically, students will jump to 60 or 62 as the age (number) from which they can find all three ages. If a student makes a guess of 58, 60, and 62, the student, by checking that guess, will quickly see that his answers are a bit low. He will soon arrive at the correct answers of 60, 62, and 64. (The equation $x + x + 2 + x + 4 = 186$ works as well.)

(2) *Make a sketch/picture.* Re-read the Careless Cat problem, #12. Many students choose to solve it by actually sketching a six-foot deep hole, then tracing the cat's up-and-down climb. Frequent student use of this technique provides the teacher with a clue that the student may be a visual learner, and that he/she may have strength in the visual/spatial intelligence area.

Making a sketch/picture is not to be seen as a technique inferior to an algebraic approach, or to other algorithmic approaches.

(3) *Make a table or chart.* This method is especially useful when the problem involves trends. Producing a table or chart may help to make that trend more clear to students. This problem is easily solved by making a simple table: "For one week beginning January 14, the daily high temperature went up by 3 degrees from the day before. What was the high temperature on January 21?"

(4) *Act it out.* For some, distilling a problem to its essentials on paper is difficult, and they must, if possible, literally act out the problem. Kinesthetic learners, in particular, may need to have access to this technique in the classroom. Here is a well-known problem that is well-solved by making a sketch or by acting it out: "Five strangers meet at once. Each shakes hands with all the others once. How many handshakes will there be?"

(5) *Look for patterns.* This technique encourages students to ask, "What's happening here?" Let's look at this problem: "What's the next number in this series? 1, 4, 9, 16, 25, ___ ?" Although one way of looking at the pattern is to identify it as "squares of the real numbers," it can also be described this way: "The numbers are separated by consecutive odd numbers." In fact, that second way of describing the series is the one more likely to be offered at first crack by middle schoolers.

Looking for patterns is especially useful in geometry. In fact, use of the phrase "Do you sense or see a pattern here?" should become standard teacher-talk in mathematics classrooms.

(6) *Make a list.* The census taker challenge, #8: That's Not Enough Information! Part 1 is a good example of a problem that all but requires that one make a list (in that case, a list of factor trios that give 72).

Making lists can clarify the situation and provide opportunities for elimination of improbable or wrong answers.

(7) *Work a simpler problem.* Look back at the "five strangers meet" handshakes problem above, in the "act it out" section. What if the original problem had referred to *fifteen* strangers meeting? Acting it out may have been impractical, and a sketch may have gotten too messy to be worthwhile. By solving simpler but similar problems (three strangers? four?), students may see patterns or shortcuts which will allow them to solve the more difficult one. For instance, some experimentation with the "handshakes" problems will reveal that the formula $[n(n-1)]/2$ will always give the number of handshakes (where n is the number of strangers meeting).

(8) *Use "if/then."* The classic "Farmer at the River" problem is typically solved with many "if/thens" being uttered: "A farmer wishes to cross a river with his bag of corn, his chicken, and his pet fox. The leaky small boat available for crossing will only hold the weight of the farmer and one other thing (the corn, the chicken, or the fox). Naturally, the farmer must make several trips back and forth across the river. The problem is, if the chicken is left alone with the corn, it will eat it, and if the fox is left alone with the chicken, the chicken will be eaten! (*Note:* The fox does not care for corn.) How can the farmer get himself and the corn, chicken, and fox across the river?" It is difficult to imagine this problem being solved without "if/thens!"

(9) *Solve an equation.* Transforming a problem situation into an equation to be solved is, to some extent, what algebra is all about. Knowledge of, and the ability to use algebra, are invaluable; it should be acknowledged, however, that many students will "see" the elegance of algebra more quickly than others, and that the use of the other problem-solving strategies should never be discounted as inferior to algebra.

APPENDIX B

Multiple Intelligences Theory

OVER THE LAST twenty years, Harvard University educator Howard Gardner's Theory of Multiple Intelligences has prompted many teachers to re-examine their instructional practices. Although Gardner provides a fairly complex technical definition of intelligence (much of it tied to his clinical research with brain-damaged patients), he does summarize by saying that intelligence is best defined as the ability to solve problems. He further claims that traditional IQ tests ignore many other intelligence strengths, to the detriment of children and their learning.

Gardner claims that he has discovered at least seven distinct intelligences that fit his definition. They are:

- verbal/linguistic, defined as sensitivity to the meaning of words
- mathematical/logical, defined as highly developed logical and reasoning abilities, and the ability to recognize patterns
- bodily/kinesthetic, defined as the ability to use the body in a skillful manner
- musical, defined as sensitivity to melody and rhythm
- spatial, defined as the ability to perceive the world in three dimensions and to transform objects and space mentally
- interpersonal, being the ability to understand others well
- intrapersonal, being the ability to understand oneself well

The greatest impact of Multiple Intelligences Theory may be that it has helped provide teachers with an alternative way of thinking about student learning. It is becoming increasingly difficult for teachers to justify uttering the phrase "But we covered that!" when students do not demonstrate understanding. Fortunately, Gardner's work, and the work of others who see wisdom in his theory, have provided teachers with new and powerfully effective techniques for reaching all students.

There are now many books on the topic of multiple intelligences, and on how the theory can be put into practice in the classroom. For teachers who are just beginning their investigation into this, however, there is one key idea to keep in mind. Reference to this single theme alone may dramatically increase the effectiveness of teachers' lessons. This is it: When planning lessons and units, teachers should be sure that presentations, assignments and other activities, and assessment options take into account students' multiple intelligence strengths and weaknesses.

EXAMPLES

When teaching about the U.S. Civil War, the teacher may want to provide recorded examples of music from the period. This will likely help to engage and motivate those with strong musical intelligence in the classroom.

When given the homework assignment of finding out about the state of Florida, students can be given several options: Produce a short radio talk show interview skit wherein the "governor" of the state chats with listeners, or produce a large, colorful poster of the state, with points of interest noted by drawings or photographs cut from magazines, etc.

When it is time to assess student understanding of a lesson or a unit, the teacher might give students various assessment options, allowing students to use their stronger intelligence area. Composing silly (but accurate) poems that present various geometric formulas is one idea—and it *does* work!

A special benefit of referring to this theory for the mathematics teacher is that it can free the teacher and students from slavish adherence to strictly mathematical/logical themes—the theory shows that there is not only *room* for the other intelligences in the math classroom, it *demands* that they be made full partners in the learning process.

APPENDIX C

Eighteen Ideas for Middle-Grade Mathematics Classrooms (a Potpourri)

(*1*) MAKE SURE THAT the classroom has adequate supplies to support an innovative and progressive middle-grade program, one that is in tune with the NCTM Standards. While there can be no definitive list of supplies, here are some suggestions. The teacher who has read this book will recognize many as essential to presentation of some of the book's problems.

Classroom sets:

geoboards
unit cubes, tens rods, hundreds flats, thousands cube
calculators
solid plastic English/metric rulers
flexible tape measure, English/metric
cardboard, or better yet, plastic geometry templates
protractors
compasses (well-made! It is of no value to purchase cheap metal compasses.)
six-sided dice
"alternative-sided" dice, eight- or ten-sided
think pads (stapled half-sized pads of scrap paper)
thousands of counters (bingo chips, etc.)
graph paper
colored pencils
scissors, tape, glue

In addition, each middle-grade mathematics classroom should have the following items available/handy/posted:

a geometric solids set

a plastic transparent volume set
an overhead projector calculator
a large number line with positive and negative numbers noted
a variety of relevant posters (pi to one hundred places in large print, the first one thousand digits of pi, polygon identification posters, etc.)
a full-sized cubic meter representation (made of fit-together plastic pipes)
several stop watches
dozens of "sponge" activity opportunities [more on this below; see idea (2)]

Many of these items can be student-produced. A "pi-to-one-hundred-places" poster, for instance, can be produced by a student or small team of students.

(2) Provide many "sponge" activities for students. One idea is to purchase several dozen hands-on math and logic-related puzzles, à la Rubik's Cube (self-contained is best—avoid loose pieces) and to put them individually in empty shoeboxes. These can be readily available for students without restriction, or available at designated times only. The teacher may want to offer small rewards for any students who can solve the more challenging puzzles. Allowing him or her to explain the solution to the rest of the class is also a good idea.

Some of these puzzles can be costly, but their use in middle-grade math classrooms can be highly motivational, therefore, worthwhile. Funding possibilities include district-level offices (a mathematics office?), building PTA or PTOs, and generous parents or area businesses. Where there is a will, there is a way.

(3) Another "sponge" idea: Provide students boxes full of blank tic-tac-toe game sheets (these can be duplicated on the back of scrap paper). There are several other simple strategy games that can also be put onto such sheets; the famed "Brussels Sprouts" is one. (Brussels Sprouts is reported by famous math puzzler Martin Gardner to have been invented jointly by John Horton Conway and Michael Stewart Paterson of Cambridge.) Requiring a little more effort: Providing copies of challenging mazes for students to attempt during designated "sponge" times.

(4) Teach students how to take their own pulse rates. Having this skill opens up the doors to many math classroom activities (production of graphs, discussion of why individual and class average rates vary,

etc.). It will also win the appreciation of the Physical Education instructor, as it is an excellent interdisciplinary activity.

(5) Offer *basic facts* race opportunities. Occasionally, critics of the NCTM Standards will imply that those Standards discourage teachers from asking students to memorize basic addition, subtraction, multiplication, and division facts. This is in error. Students need to know these basic facts, and they may be appropriately reinforced even at the middle grades. One way to do this is to challenge students to beat the previous best classroom time for orally giving the answers to, say, 50 basic facts problems. The teacher might verbally present a mix of problems (for example "7 × 13"), while a student keeps track of the time. The participating student might be eliminated if even one answer is incorrect, or the teacher may choose other format options. "Best times" can be permanently posted in the classroom, perhaps keeping track of different sections of students separately. A chart might look like this:

Name	Date	Time
~~Sylvia W.~~	~~3/24~~	~~47.4 seconds~~
~~Ted C.~~	~~3/25~~	~~47.1 seconds~~
Ann P.	3/27	46.2 seconds

A "championship" round, with various classroom champs participating, could be held at lunchtime one day. Naturally, the tone of these activities should emphasize fun and participation.

(6) Conduct a "Daily Problem" contest. This is a good way to open math class; it is a classroom behavior management tool as well. One format: The teacher presents a single math or logic problem or puzzle to the entire class; everyone has two minutes to individually turn in a solution to the teacher (without names on the slips of paper). The teacher may want to challenge the class to beat the 50% correct mark.

(7) Occasionally provide students with "math/logic trivia quizzes"; classic just-for-fun questions with a twist. These can be class warm-ups or optional evening activities for interested students. Some sample questions, with answers provided:

    a. Do they have a Fourth of July in England? (Of course they do; it's the day after the third of July.)

    b. Why is it illegal for a woman living in Washington D.C. to be buried west of the Mississippi River? (Because she's alive.)

c. If you take two apples from three apples, what do you have? (*You*, having taken two apples, *have* two apples.)
d. Farmer McMullen had 19 sheep; all but seven died. How many did he have left? (Seven, as it says.)
e. Multiply the first twelve whole numbers together. What is the product? (Zero, which is the product *any time* one of the factors is zero.)
f. If an archaeologist claims to have found a coin with the date 92 B.C. stamped on it, should he be believed? (No, "B.C." was not a convention then.)
g. How many birthdays does the average U.S. male have? (One; the others are really birthday anniversaries.)
h. Someone builds a square house, each side faces south. If a bear strolls by, what color is it? (It's a white polar bear. The only place one can build a house with four sides facing south is directly on the north pole.)
i. How many twenty-cent stamps in one dozen? (Twelve, as there always are with one dozen.)

(*8*) Conduct math scavenger hunts with students. There are many possible formats for such a hunt; they might be limited to finding "items" in newspapers, or within the classroom, or the school, or at home. They might work alone, or with partners. Certainly, these hunts should be enrichment activities, and students' success (or lack thereof) at assembling the items should not be reflected in their math class assessment.

Here is a sample list of "items" that students might need to find in the newspaper:

a. tomorrow's expected high temperature in Cincinnati
b. the name of a professional sports team whose current winning percentage for the season is within 10 points (ten thousandths) of .585
c. a number between one billion and ten billion
d. an advertisement promoting an item costing exactly $7.99
e. an advertisement promoting items at 40% off the regular price
f. a statement of distance between ten and twenty miles
g. yesterday's Dow Jones average
h. a reference to a certain number of animals
i. a reference to the weight of something between one hundred and two hundred pounds

The teacher may insist that students bring in the actual reference from the newspaper, or, they might be required to simply note exactly where the item came from (name of newspaper, date, page, etc.).

Here is a sample list of "items" with no limits on the source:

a. a reference in a recipe that calls for two cups of flour
b. an estimate, along with a rationale, for the number of books in the school media center
c. the height of the Taj Mahal
d. three examples of ovals within the school grounds
e. the zip code for Mt. Rushmore
f. from a food nutritional label, a serving of some food which supplies 10% of the RDA of sodium
g. the name of a famous person born in 1946
h. a real telephone number containing five 5s
i. the area, in square miles, of the world's second largest country
j. a catalog item costing more than $10,000

(9) Make occasional, planned, public, horrible mathematical mistakes, hoping that one or more students will catch your error. If one does, the teacher should be prepared with some clever recognition ritual; the awarding of a "Caught-the-Teacher" certificate, for instance. An option is for the teacher to purposefully make an "error of the day." This may motivate students to pay more careful attention to the teacher's presentations! If the teacher chooses this "one-a-day" format, and if no student spots it, he or she may want to announce the missed mistake at the end of each class period.

(10) Have students write mathematics and education-related letters to newspaper editors. This will encourage students to look in the newspaper each day. Topics can be as straightforward as "here's what we're learning in math class these days." Many newspaper editors are soft touches when it comes to publishing student-produced letters. Other students and/or the teacher should always carefully proofread any letters that might be sent; editors may print them exactly as submitted.

(11) Plan a Parents' Math Night. Although this takes a great deal of planning, it can go a long way toward garnering parental support for a teacher's and a school's mathematics program. Keys are engaging parents in some lively activities, and in avoiding turning the event

into a simple listing of topics covered in math class (deadly!). Students should definitely be integral parts of this endeavor.

In many homes, parents unwittingly subvert math teachers' efforts with comments such as "I never liked math," or "My daughter doesn't do well at mathematics, but she comes by it honestly." A Parents' Math Night provides an opportunity to discuss this unfortunate phenomenon of adult disparagement of mathematics.

(12) Invite speakers into the math classroom. Who might be appropriate? Bankers, insurance agents, and others in the field of finance might be good to ask; often, those kinds of companies have packaged presentations ready. (Beware the overly commercial versions.) Local colleges and universities are also a good source for speakers. It can't hurt to write letters to mathbook authors, inviting them to the school.

(13) Use video clips in the mathematics classroom when possible. If one keeps one's eyes open, one will find many short clips that are relevant. The "7 × 13 = 28" problem, #52, is from an old *Abbott and Costello* television show; it is available on video. Video recordings of TV commercials make great fodder for classroom discussion, as do recordings of sports score updates. Middle-grade students (and most all modern American children) tend to automatically look at a television screen if it is turned on; teachers may as well make the most of the phenomenon.

(14) Conduct Rapid Rabbit Relays. These are low-stakes contests conducted, typically, among teams of four students each. Each team is put into a straight row, one student behind the next. The first student in each team is given a sheet of paper with four different math problems on it; his or her job is to quickly solve the first problem, then to pass the sheet to the next teammate, who solves the next problem, and so on to the fourth person. The winning team is the one which finishes first, of course. The twist here is that each new student needs the solution from the previous problem to solve his or her problem. Thus, a Rapid Rabbit contest sheet might look like this:

First Team Member: Circle the only prime number in this set:

9   15   21   37   49

Eighteen Ideas for Middle-Grade Mathematics Classrooms (a Potpourri) **123**

Second Team Member: Add 83 to the circled number above; put the sum in this blank:

———

Third Team Member: Divide the number above by 6; put the quotient in this blank:

———

Fourth Team Member: List all of the factors of the number above.

Naturally, the team must have the last answer correct to be declared the winner. The teacher may want to consider permanently posting the fastest Rapid Rabbit times. The activity can even be expanded into a cross-class section contest.

(15) Conduct "Quick Sketch" contests at the chalkboard. An example: Have all of the students in the classroom line up and, one at a time, draw on the chalkboard their best circle, or best perfect square. Each student should be given only one chance; whatever each produces stays on the board. When all are done, and are seated, the teacher can rate each attempt on the board by assigning each drawing a number from one-to-ten, with ten being perfect. The student responsible for the best drawing might be entitled to a small reward. Naturally, this sort of contest needs to kept lighthearted. Other suggestions for drawings and sketches:
  a. a 45 degree angle (or any other specified number of degrees)
  b. a rectangle which is twice as long as it is wide
  c. an equilateral triangle
  d. a brace {
  e. a segment exactly one foot in length (or any given length)

  A benefit of this activity is that it allows for student movement [see idea *(18)* below].

(16) Incorporate humor into the classroom by way of displaying cartoons (there *are* math-related comic panels out there!) and by sharing "dumb sayings." Actual "dumb sayings," said by real people, include:

"I want to gain 1,500 or 2,000 yards, whichever comes first." *(from George Sanders, former New Orleans Saints running back)*

"Half this game is 90% mental." (*from Danny Ozark, former Major League manager*)

"Baseball is 90% mental. The other half is physical." (*from baseball great Yogi Berra, who apparently went to the same school as Danny Ozark*)

"It's about 90% strength and 40% technique." (*from World Middleweight Wrist-Wrestling champ Johnny Walker, on what it takes to be a champion. Mr. Walker also apparently attended school with Mr. Ozark and Mr. Zimmer.*)

"It could just as easily have gone the other way." (*from Chicago Cubs manager Don Zimmer, referring to his team's 4-4 record on a road trip*)

"If crime went down 100%, it would still be fifty times higher than it should be." (*from Washington D.C. Councilman John Bowman*)

The occasional silly riddle or joke will also help to enliven any math class. Any groans from middle-grade students after hearing these should only serve to confirm that they hit their intended marks.

Riddle: If two's company, and three's a crowd, what's four and five? (Nine, of course.)

Joke: A man was selling his car by way of a classified ad. Needing to leave his house one evening, he instructed his twelve-year-old daughter to handle any telephone calls which might come in regarding the car. A call came in, and the girl answered. When the caller asked "How much is your dad asking for the car?" the girl responded, "Well, he's asking $3,000, but won't take anything under $1,800."

(*17*) Explore with students the work of Dutch artist Maurits C. (M.C.) Escher. His art is always fascinating to middle-grade students. His themes of mystery, infinity, absurdity; his use of symmetry, tessellation, and "impossible figures" are irresistible to adolescents. There are numerous Escher posters available, many books, even T-shirts, neckties, etc.

(*18*) Allow, if at all possible, for student movement, for drinks of water, and for appropriate snacks, especially if the class period is a long one. There are many easy and non-disruptive ways for teachers to incorporate these things into the classroom routine; students will be appreciative, *and* more attentive.

INDEX

Abbott and Costello, 122
Abundant numbers, 60
Adjacent sides, 72
"Acting it out" (problem solving strategy), 27, 96, 112
Active participation, 11
Adolescents, characteristics of, 4–6
Algebra, 25, 26, 27, 77, 88, 111, 112, 113
Algorithms, 79–82, 103, 112
Angles, 72, 123
Anticipatory set, 12
Archimedes, 97
Area, 30–32
Arithmetic, 79
Assessment, 3, 10, 18, 43, 82, 103, 116, 119, 120
Attention, student need for, 5, 9
Average (mean), 22

Balance (banking), 100
Banking, bankers, 22, 122
Berra, Yogi, 124
Bodily/kinesthetic intelligence, 115
Bowman, John, 124
Brainstorming, xi, 26, 62, 96
"Brussels Sprouts" game, 118
"Bumblebee paradox," 96

Calculators, 18, 24, 67, 76, 98, 100, 107, 109
Calendar, 40, 41–42, 48
Charts, 119
Chih, Ch'ung, 98
Circles, 32–33, 33–34, 38, 97–99
Circumference, 33–34, 97, 98
Closure, 16–17

Combinations, 101, 103
Competition, 83
"Common" numbers, 84
Composite numbers, 61
Compound interest, 22–24, 99–100
Computation, 111
Computers, 98
Conferencing, 3
Confidentiality, 7
Connections, 82, 85
Consumer mathematics, 22, 25, 99
Contests, 44, 89, 119
Conway, John Horton, 118
Counterfeit money, 73
Counterintuitive problems, 2, 4, 21–55
"Covering" material, 17, 18
Curves, 38
Cycloid, 39

"Daily Problem" contest, 119
Dead time, 16
Debate, 3, 4
Decimals, 24, 46–48, 97, 107
Deficient numbers, 60
Density, 88
Diagonals, 65, 72
Diameter, 33–34, 98
Dice, 107–110
Differences, 30
Digits, 93
Discipline, student, 2, 8, 16
Discount, 25
Discussion, 2, 3, 4, 18, 62, 91, 93, 95, 96, 104, 118
Doubling, 76–77
Drill, 3

127

128 Index

Effective teaching, 6–18
Engagement, student, 1, 2, 18
Equations, 113
Equations, solving (problem solving technique), 113
Escher, M.C., 124
Estimation, 43, 44–45, 100–101
Euclid, 61
Extra sensory perception (ESP), 105–107
Eye contact, 8

Factors, 28, 60–61, 112
Fairness, 6, 14
Flight of the Bumblebee, The, 103
Forgetting, 17
Formulas, 103
Fractions, 21, 37–38, 46–48, 107–110
"Free time," 16
Functions, 38, 61

Games, 79–93
Gardner, Dr. Howard, 115
Gardner, Martin, 118
Geometric growth, 77
Geometry, 30, 32, 38, 42, 61, 63, 65, 69, 71, 95–97, 112
Gershwin, George, 101–103
Grades, grading: *see* Assessment
Graphs, 118
Group instruction, 14, 15
Guess and check (problem solving technique), 111–112
Guessing, 42–44, 44–46, 106

Heartbeats, 101, 118
Hiram the Builder, 97
HIV virus, 89–91
Homework, 3, 13, 18, 71
Humor, 17, 123–124
Hungarian Rhapsody, 103

"If/then" (problem solving technique), 113
Inference, 86–87
Infinity, 47, 87, 95–97, 124
Insurance agents, 122
Integers, 35–36

Interdisciplinary approach, 2, 18, 33
Interest, 22, 24, 99–100
Interpersonal intelligence, 115
Intrapersonal intelligence, 115
IQ tests, 115

Johns Hopkins University, 2

Kinesthetic learners, 15, 63, 95, 107, 112
Kings, biblical *Book of,* 97

LaVoie, Dr. Richard, 14
Language arts, 2
Length, 42–44
Lesson plans, 116
Level of concern, 10, 14, 17
List, make a (problem solving technique), 112
Listening, 7
Loans, 99
Logic, 25, 35, 36, 37, 44, 46, 48, 54, 57, 59, 63, 76, 95, 103–105, 119

Manipulatives, 64
Markup, 25
Materials, 117
Mathematical/logical intelligence, 115, 116
Mazes, 118
Measurement, 42–45
Memorization, 68, 101–103
Mental arithmetic, 67–69, 75, 91–92, 92–93
Millennium, 48
Misbehavior, student, 5, 6, 7
Mnemonic devices, 99
Money, 70, 73
Monopoly, 82, 85
Moonlight Sonata, 103
Mortgages, 99–100
Motivation, student, 6, 9, 10, 13, 118
Movement, student, 14, 15, 16, 123, 125
Multiple Intelligences Theory, 115–116
Music, 101–103
Musical intelligence, 115, 116

National Council of Teachers of Mathematics (NCTM), 2, 3

National Middle School Association, 1
Newspapers, 120–121
Number lines, 87
Number sense: *see* Numeracy
Numeracy, xi, xii, 18

Odd numbers, 93
Odds, 51, 99, 106
Ozark, Danny, 124

Palindromes, 77–78
Parade, 50, 53
Parallelograms, 72
Parents, 121–122
Paths, 65–67, 69–70
Patience, 17
Patterns, 27, 62, 98, 112, 113
Patterns, looking for (problem solving technique), 112
Patterson, Michael Steward, 118
Paulos, Dr. John, xi
Percent, 21–23, 25, 59, 99, 103–105
Perfect numbers, 60
Phenolphthalein, 89–91
Pi, 32–33, 33–34, 97–99, 118
Polygon, 72
Popular culture, 9
Portfolios, 3
Praise, 8, 12
Predicting, 76
Prime factorization, 47
Prime numbers, 61
Prior knowledge, 12
Probability, 27, 40, 46, 50, 53, 89, 107–110
Problem solving, 2, 3, 18, 19, 26, 34, 37, 49, 51, 111–113
Products, 28–30, 91
PTA, PTO, 118
Pulse rate: *see* Heartbeats
Puzzles, 57–78, 118

Quadrilaterals, 72
Questioning technique, 9, 11–14, 15
Quizzes: *see* Assessment
Quotients, 30

"Rapid Rabbit Relays," 122

Rasch, Dr. Kathe, 1
Rate, 34, 37, 58–59, 118
Ratio, 45, 97, 98
Rationale, 17, 44, 76
Rectangles, 65, 66, 71, 72, 123
Reflection, student, 4
Rhapsody in Blue, 101–103
Rhombi, 72
Risk-taking, 6, 7
Rounding, 109
Rubik's Cube, 118
Rules, formulating, 62

Sanders, George, 123
Sarcasm, 7, 8
Scavenger hunts, 120
Segments, 69
Self-esteem, 3
Sequencing, 27, 61–62, 77, 98, 102–103
Silent lesson, 14
Sketch, making a (problem solving strategy), 34, 37, 49, 112
Slope, 88
Sodium hydroxide, 89–91
Solitaire, 63
Spatial discrimination, 72
Spatial intelligence, 115
Speakers, classroom, 122
Sponge activities, 16, 69, 72, 118
Square roots, 67–69
Squares (figures), 72
Squaring, 76, 78
Standards, National Council of Teachers of Mathematics Curriculum and Evaluation, 2, 117, 119
Statistics, 27
Status, student need for, 5
Subtraction, 91
Sums, 28, 30
Surveys, 15
Symmetry, 124

Table, make a (problem solving technique), 112
Tables, 107–110
Tax, 25, 26
Temperature, 44
Test: *see* Assessments
Tic-tac-toe, 118

"Ticket out", 14
Time, 26, 35, 37, 39, 59, 71
Touch, 8
Trapezoid, 72
Trends, 112
Triangles, 63–64, 123
Trial and error (problem solving strategy), 26
Trigonometry, 88
Trust, 7

"Uncommon" numbers, 84
Understanding, checking for, 13

Verbal/linguistic intelligence, 115
Vertex, 72

Visual learners, 112
Volume, 44–45, 76, 88–89
Von Leibnitz, Gottfried, 98
Vos Savant, Dr. Marilyn, 50–53, 53–54

Wait time, 11, 12
Walker, Johnny, 124
Weight, 44, 73–74, 88–89
Whole numbers, 91
Work a simpler problem (problem solving technique), 113
Writing, 3, 18

Zeno of Elea, 95
Zimmer, Don, 124

ABOUT THE AUTHOR

NELSON JOHN MAYLONE is a doctoral Fellow at Eastern Michigan University, has taught elementary and middle grades mathematics methods at the University of Michigan, Dearborn, and teaches in the Graduate School of Education of Michigan State University. He has been a middle school and elementary school administrator, and was a middle-grade mathematics educator for sixteen years.